Developments in Geotechnical Engineering 18

AF815685

STRESS IN SUBSOIL AND METHODS OF FINAL SETTLEMENT CALCULATION

Czechoslovak Academy of Sciences

Further titles in this series:

1. *G. SANGLERAT*
THE PENETROMETER AND SOIL EXPLORATION

2. *Q. ZÁRUBA AND V. MENCL*
LANDSLIDES AND THEIR CONTROL

3. *E. E. WAHLSTROM*
TUNNELING IN ROCK

4. *R. SILVESTER*
COASTAL ENGINEERING, I and II

5. *R. N. YOUNG AND B. P. WARKENTIN*
SOIL PROPERTIES AND BEHAVIOUR

6. *E. E. WAHLSTROM*
DAMS, DAM FOUNDATIONS, AND RESERVOIR SITES

7. *W. F. CHEN*
LIMIT ANALYSIS AND SOIL PLASTICITY

8. *L. N. PERSEN*
ROCK DYNAMICS AND GEOPHYSICAL EXPLORATION
Introduction to Stress Waves in Rocks

9. *M. D. GIDIGASU*
LATERITE SOIL ENGINEERING

10. *Q. ZÁRUBA AND V. MENCL*
ENGINEERING GEOLOGY

11. *H. K. GUPTA AND B. K. RASTOGI*
DAMS AND EARTHQUAKES

12. *F. H. CHEN*
FOUNDATIONS ON EXPANSIVE SOILS

13. *L. HOBST AND J. ZAJIC*
ANCHORING IN ROCK

14. *B. VOIGT (Editor)*
ROCKSLIDES AND AVALANCHES, 1 and 2

15. *C. LOMNITZ AND E. ROSENBLUETH*
SEISMIC RISK AND ENGINEERING DECISIONS

16*A. C. A. BAAR*
APPLIED SALT-ROCK MECHANICS, 1
The in-situ Behavior of Salt Rocks

17. *A. P. S. SELVADURAI*
ELASTIC ANALYSIS OF SOIL-FOUNDATION INTERACTIONS

Developments in Geotechnical Engineering 18

STRESS IN SUBSOIL AND METHODS OF FINAL SETTLEMENT CALCULATION

by

JAROSLAV FEDA

Institute of Theoretical and Applied Mechanics
of the Czechoslovak Academy of Sciences, Prague

ELSEVIER SCIENTIFIC PUBLISHING COMPANY

Amsterdam • Oxford • New York • 1978

Scientific Editor
Prof. Ing. Dr. Zdeněk Bažant, DrSc.

Reviewer
Doc. Ing. Ján Jesenák, CSc.

Published in co-edition with
ACADEMIA, Publishing House of the Czechoslovak Academy of Sciences, Prague

Distribution of this book is being handled by the following publishers

for the U.S.A. and Canada
Elsevier/North-Holland, Inc.
52 Vanderbilt Avenue
New York, New York 10017

for the East European Countries, China, Northern Korea, Cuba, Vietnam and Mongolia
Academia, Publishing House of the Czechoslovak Academy of Sciences, Prague,

for all remaining areas
Elsevier Scientific Publishing Company
335 Jan van Galenstraat
P.O.Box 211, Amsterdam, The Netherlands

Library of Congress Cataloging in Publication Data

Feda, Jaroslav.
Stress in subsoil and methods of final settlement calculation
(Developments in geotechnical engineering; 18)
Updated translation of Napjatost podzákladí a metody výpočtu konečného sednutí.
Bibliography: p.
Includes indexes.
1. Soil mechanics — Mathematical models. 2. Soil consolidation — Mathematical models. I. Bažant, Zdeněk.
II. Title. III. Series.
TA 710·F 4313 624'·1513 78—1612
ISBN 0—444—99800—4
ISBN 0—444—41662—5 (Series)

© *J. Feda, Prague 1978*

All rights reserved. No part of this publication may be reproduced, stored in a retrieval system, or transmitted in any form or by any means, electronic, mechanical, photocopying, recording, or otherwise, without the prior written permission of the publishers.

Printed in Czechoslovakia

CONTENTS

LIST OF SYMBOLS

a, a_0	constants, parameters
a_v	modulus of compressibility
A, A_s	coefficient of pore pressure
A_0	coefficient of compressibility
b	a constant, a parameter
B	width of a rectangular loading area
	– side of a square loading area
	– coefficient of pore pressure
B_0	least dimension of a loading plate (= 0.3 m)
c', c	effective cohesion, cohesion
c_u	undrained shear strength
C, C_0	coefficient of compressibility
C_c	compression index
C_p	cone penetration resistance (in static penetration)
d, d_0, d_z	depth
	– a parameter
d_{act}	depth of active zone
D	diameter (of circular loading plate, of pressure cell)
D_0	diameter of earth pressure cell diaphragm
e	void ratio
e_0	initial void ratio
E	Young's modulus
E_h, E_v	Young's modulus in horizontal, vertical direction
E_0	Young's modulus for $z = 0$ or $\sigma_{oct} = 0$
E_i	modulus of deformation
E_{oed}	oedometric modulus of deformation
E_c	modulus of elasticity of an earth pressure cell
E_p	pressiometric modulus of deformation
E_t	tangent modulus of deformation
E_s	subsoil modulus of deformation
E_u	undrained modulus of deformation
E'	modulus of elasticity in system of effective stresses

f	a parameter
F	a constant
g	a parameter
G_i, G, G_0	modulus of shear deformation (modulus of elasticity in shear)
h	depth of the point of action of a force
h_i, Δh_i	thickness of an elementary layer in the subsoil
H	concentrated horizontal force
I, I_w, I'_w	coefficients for settlement calculation
I_p	plasticity index (%)
k	an exponent, a parameter
k_s	modulus of vertical subgrade reaction
K_0 (K'_0)	coefficient of lateral pressure at rest (in effective stresses)
K_0^n	coefficient of lateral pressure at rest of normally consolidated soil
K'	ratio between effective stresses σ_h and σ_v
K, K'	coefficient (with different subscripts) for calculation of stresses in elastic halfspace
L_2, L'_2	coefficient for calculation of stresses in elastic halfspace
lg, log	natural, common logarithm
m	an exponent
m_v	modulus of volume compressibility
M, M'	coefficient (with different subscripts) for calculation of stresses in elastic halfspace
MPa	unit of stress (1 MPa = 1 . 10^3 kPa = 1 . 10^6 N/m^2 $\doteq$ 10 kg/cm^2)
n	an exponent
$n = E_h/E_v$	ratio of Young's (deformation) moduli
n_0	initial porosity
N	standard penetration resistance
N_2, N'_2	coefficient for calculation of stresses in elastic halfspace
OCR	overconsolidation ratio
p	a component of a linearly distributed load of halfspace
p_a	atmospheric pressure
p_k	capillary pressure
p_l, p_z	conjugate stresses in a slope
P	concentrated vertical force
q	uniform vertical load – a half of the difference ($\sigma_v - \sigma_h$)
r	a cylindrical coordinate (z, r, Θ)
R, R_1, R_2	radius vectors of the point of action of a load
R_f	ratio between real and asymptotic difference of principal stresses
s	a component of a linearly distributed load of halfspace
t	a component of a linearly distributed load of halfspace
t_0	diaphragm thickness

u	neutral (pore) pressure
v	height (thickness) of an earth pressure cell or sample
w, w_{oed}, w_T	vertical, oedometric, total vertical displacement (settlement)
w_0	settlement of loading plate, settlement of foundation centre
x, y, z	rectangular (Cartesian) coordinates and axes
α	coefficient of pore pressure
	– angle between the normal to contact surface and the horizontal
α_1, α_2, α_0	coefficients, parameters
β	slope inclination
β_0	coefficient
γ_d	unit weight of dry soil
γ_n	unit weight of water saturated soil
γ_{oct}	octahedric shear strain
γ_{xy}, γ_{yz}, γ_{zy}	shear strains in the x, y, z directions
γ_s	specific weight of soil grains (solid phase)
γ_w	specific weight of water
γ_0	unit weight of soil with natural water content
γ'	unit weight of submerged soil
δ_0	relative density of soil with natural water content
δ_s	relative density of soil grains (solid phase)
ε	coefficient for calculation of stresses in elastic halfspace
ε_h, ε_v	horizontal, vertical normal strain (compression +)
ε_i	strain intensity
ε_{vc}	axial strain of sample in consolidation phase
ε_{vol}	volume strain of sample in oedometer
ε_x, ε_y, ε_z	normal strains in the x, y, z directions
ε_i^e, ε_i^p	reversible, irreversible component ($i = 1, 2, 3$ or a, r) of strain
ε_a, ε_r	axial, radial strain
ε_1, ε_2, ε_3	principal strains
η	coefficient of lateral pressure
Θ	angle – cylindrical coordinate (z, r, Θ)
λ	model scale
λ_1, λ_2	shape factors
μ, μ_1, μ_2, μ_3, μ_t	Poisson's ratios
μ'	Poisson's ratio in system of effective stresses
ν	concentration factor
σ_h, σ_v (σ_h', σ_v')	horizontal, vertical normal (effective) stress (pressure +)
σ_a', σ_r'	axial, radial effective stress
σ_i	stress intensity
σ_n	normal stress in a given plane (pressure +)
σ_{oct}	octahedric normal stress (pressure +)
σ_x, σ_y, σ_z	normal stresses in the x, y, z directions (pressure +)

σ_R	normal stress in the R direction
$\sigma_z, \sigma_r, \sigma_\Theta$	normal stresses in the directions of cylindrical coordinates (z, r, Θ) (pressures +)
$\sigma_1, \sigma_2, \sigma_3$	principal stresses (pressure +)
τ	shear stress in a given plane
τ_{oct}	octahedric shear stress
$\tau_{r\Theta}, \tau_{rz}, \tau_{\Theta z}$	shear stresses in cylindrical coordinates (z, r, Θ)
$\tau_{xy}, \tau_{xz}, \tau_{yz}$	shear stresses in the directions of the x, y, z axes
φ	angle between radius vector R and the vertical (axis z)
Φ_f', Φ_r'	peak, residual effective angle of internal friction
Φ_μ	angle of intergranular friction (friction between grains)

INTRODUCTION

More than forty years ago, Press (1930) published the results concerning his in situ load tests of sands and loams performed with plates of various sizes (0.8 m^2 max.). Some of these results are shown in Fig. 1. Under the application of an equal, uniform load (for example, 0.4 MPa), two plates of different dimensions (for example, 200 cm^2 and 1400 cm^2 in the case of sand) can settle equally, the smaller plate pushing the subsoil out from beneath it, the larger plate compacting it. Plates of yet smaller dimensions (i.e. less than 200 cm^2 in the case being considered) sink into the subsoil.

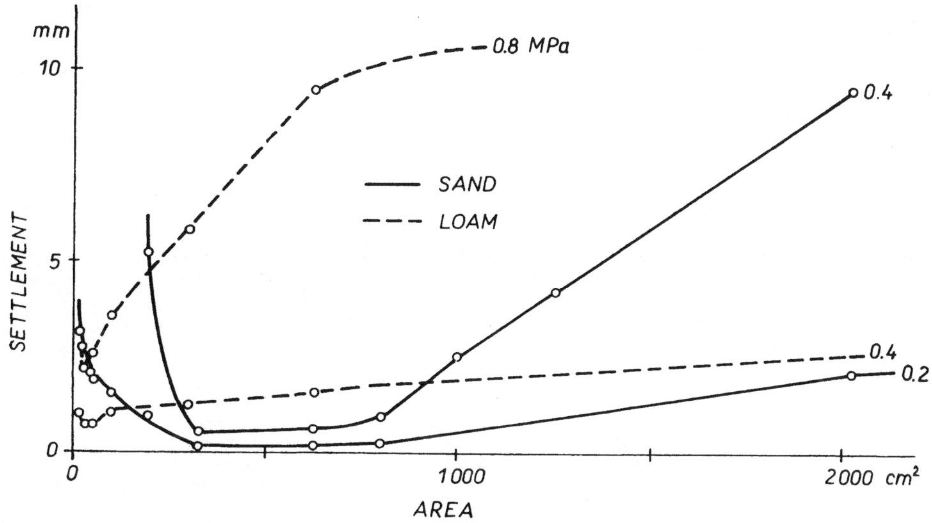

Fig. 1. Relation between settlement and area of square plates subjected to equal uniform load in sand and loam (Press, 1930).

Similar results were obtained analytically by Brinch Hansen (1966), though his conclusions are of a rather qualitative character (Sherif, 1973).

Fig. 1 is an example of the uniform conception of the permissible load of the subsoil, based on the permissible magnitude of the soil deformation (uniform settlement as well as its non-uniform component). It is a matter of secondary importance whether or not the shear strength of the subsoil is mobilized under this deformation.

The matter which is regarded as secondary from the point of view of the conception, assumes eminent importance for considerations relating to the methods of calculation, which are the only means of converting a conception into engineering reality.

Since a mathematical model for calculating deformations in an elasto-plastic half-space was devised only recently (the situation was changed a great deal by the introduction of the finite element method), the field of soil mechanics adopted a classical engineering approach to the problem of permissible load of the subsoil: after calculating the ultimate load, a value is chosen for the permissible load which is remote enough from the ultimate load (thus ensuring negligibly small plastic regions in the subsoil) and which leads to permissible displacements.

This dualism has of late reached a high degree of perfection thanks to the development of the theory of the limit states, which employs a probabilistic definition of the two fundamental limit states — the bearing capacity and the deformation. The theory of the limit states was included in the standards and codes of many countries, and its indisputable advantage is the unification of the procedures used in the design of various engineering structures of diverse materials. It can, therefore, be expected that (following some necessary refinements) it will continue to be a standard engineering method even in the future.

An essentially similar approach is adopted in this book. The writer assumes (except in the case of non-linear solutions, particularly those obtained by means of the finite element method) that the settlement is calculated for a load under which the extent of the plastic regions in the subsoil is negligible*) (otherwise he would have to exclude the most familiar methods of settlement calculation from his discussion).

The writer has tried — and let the reader judge with what success — to provide a comprehensive review of the present state of the art in the area of the calculations of the final settlement under static loads. In his considerations he takes no account of the theory of consolidation and creep (the so-called primary and secondary settlements), i.e. excludes the estimates of the time development of settlement which are discussed at length, for example, in the books by Šuklje (1969) and Zaretskiy (1967). Neither does he deal with the structure-subsoil interaction, i.e. the distribution of the contact stress, and the permissible non-uniformity of settlement. A survey of the present state of these subjects is offered, for example, by De Beer (1972), Burland and Wroth (1974), and by the monograph of Gorbunov-Posadov and Malikova (1973). The writer's approach is deterministic rather than probabilistic (cf. Resendiz and Herrera, 1969, or Folayan et al., 1970), which seems more fitting for the limit state of deformation than for the limit state of bearing capacity. And finally, the book

*) According to Davis and Poulos (1968), plastic strains begin to appear at the following factors of safety against failure: 4—8 in normally consolidated clays, 2—3 in lightly overconsolidated clays, and <2 in heavily overconsolidated clays.

considers no other kinds of settlement than those produced by a change in the state of stress, i.e. omits the effect of shrinkage, frost heaving, subsidence of loess subsoils, etc.

There were several reasons for restricting the present discussion as outlined. Some of the topics not included were recently studied in various monographs, others are still far from being fully understood (for example, the effect of structure rigidity on the deformations in the foundation base), while the omission of yet others simplifies the exposition without subtracting anything significant from its contents. And last but not least, it is believed that the comparatively better lucidity thus gained will make it possible to draw a more comprehensive picture of the future development whose many features are typical for the whole scientific discipline of soil mechanics.

The classical method of settlement calculations which the writer employs in this book is based on the knowledge of the original stresses in the subsoil, their changes produced by the process of construction, and of the deformation parameters of the subsoil. Since such data are a basic prerequisite for the settlement calculation proper, the writer divides his discussion into three chapters. The first deals with in situ stresses in the subsoil, the second with the state of stress produced in the subsoil by external loads, and the third with the methods of calculating the final settlement (including notes on the selection of deformation parameters).

Relatively least information is available on the subject of the first chapter. Of the contents of the second and third chapters the best known are the solutions of both stresses and strains obtained by the application of the elasticity theory. These solutions are reviewed in recently published comprehensive monographs of Giroud (1972, 1973), Gorbunov-Posadov and Malikova (1973), and particularly, in a book by Poulos and Davis (1974). This is the reason why, in this respect, the writer's exposition is rather elementary, and the above publications should be referred to for greater details.

A brief description of the finite element method and its application to settlement calculations is presented in a separate section of the third chapter (Section 3.8). This course of action was taken on methodological grounds, although it is factually incorrect because the finite element method is a general numerical procedure for solving boundary value problems, which has been found equally useful in solutions of stresses as of strains in continuous as well as discontinuous, linear and non-linear, isotropic and anisotropic media, etc. (cf. Desai and Abel, 1972). Hence it falls short of the criterion according to which the book is divided into its individual parts. This shortcoming is partly made up by notes appended to several sections of the book in order to point out the possibilities of the finite element method in solutions of specific problems.

Of late, the distinction between an engineering and a scientific approach to soil mechanics problems has been becoming increasingly more clearly defined. While the engineer is primarily concerned with knowing "how" to make a quantitative estimate of the behaviour of a given object (i.e. with how to select the input data and the method of calculation), the scientist studies the reason "why" a method of calculation or data selection is more successful than another under a particular set of conditions. He tries to gain an understanding of the physical nature of the processes he is dealing with, i.e., in the case being considered, of the processes which take place in the subsoil, and to devise a mathematical model capable of describing them as accurately as possible. In his attempt to pay heed to both viewpoints, the writer critically evaluates the various methods and confronts them with selected representative experimental findings.

In conclusion, the writer considers it his pleasant duty to thank Professor J.-P. Giroud for his kind permission to reproduce his tables for calculating stresses in an elastic halfspace as a supplement to this book, and to Dr. F. G. Butler for his kind permission to use his diagrams for calculating the settlement of a continuously non-homogeneous halfspace. He is thankful to the scientific editor, Professor Z. Bažant and to the reviewer, Doc. J. Jesenák, for their many helpful suggestions.

Chapter 1

IN SITU STRESSES IN THE SUBSOIL

1.1 Introduction

In situ stresses in the subsoil are stresses present in the subsoil immediately before the application of a structural load. The field of in situ stresses, which associates a definite stress tensor with each point of the subsoil, can be classified from different viewpoints. As often happens in classification, a single criterion cannot be applied consistently and convention comes into play in the ensuing compromise.

Thus, for example, Voight (1966) proposes the following grouping of so-called natural force fields (i.e. in situ stresses) according to their origin:

1 gravitational
2 tectonic: a) current
            b) residual

Among the stresses of tectonic origin, Voight also includes stresses produced by overconsolidation, i.e. gravitational stresses deriving from previous overburden a part of which has been denuded in the meantime. Another conventional feature of Voight's classification is the fact that he regards stresses induced by gravitational tectonic as tectonic stresses, which is questionable from the genetic standpoint.

In the writer's opinion it is more natural to use the criterion of mechanics and classify the in situ stresses into

1 geostatic
2 geodynamic (tectonic)

In this grouping, an element of convention can be traced back to the fact that all natural processes are dynamic and only some of them can – under certain circumstances – be considered quasi-static. The writer's classification places the overconsolidation stresses in a group with the geostatic stresses. Because they are of a specific character (only isotropic, i.e. spherical stresses, are always involved) and comparatively easy to measure, neutral stresses are discussed separately even though they can be included among either geostatic or geodynamic stresses. Particular attention is accorded to the ratio between the horizontal and the vertical normal in situ stresses, to the so-called coefficient K_0 since it is frequently an important characteristic of the in situ state of stress and is likely to influence markedly the predictions of settlement. By the same token, overconsolidation achieved by unloading the subsoil by an excavation is also regarded as a special case. From the point of view of the geological time, this is a short-time unloading caused by man's activity, which is directly related to settlement calculations.

In the text that follows, in situ stresses are mainly judged with respect to the prediction (calculation) of the settlement.

1.2 Geostatic stresses

Geostatic stresses are produced by the weight of the overburden. If the thickness of the overburden was never reduced in the past, the subsoil is normally consolidated; if the overburden was partially denuded in the past, the subsoil is overconsolidated.

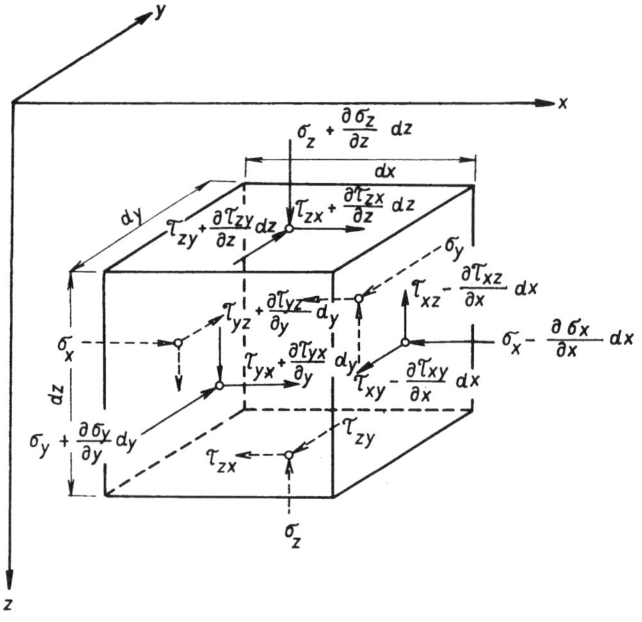

Fig. 2. Stress tensor components.

It is usually assumed that the state of geostatic stress is axi-symmetrical, with the vertical normal stress σ_z; $\sigma_x = \sigma_y$ are the horizontal normal stresses (Fig. 2). This assumption is a special case of a generally possible geostatic state of stress.

If the acceleration (except that due to gravity) is zero in the directions of the

x, y, z axes, the conditions of equilibrium require that the sum condition

$$\frac{\partial \sigma_x}{\partial x} + \frac{\partial \tau_{yx}}{\partial y} + \frac{\partial \tau_{zx}}{\partial z} = 0$$

$$\frac{\partial \sigma_y}{\partial y} + \frac{\partial \tau_{xy}}{\partial x} + \frac{\partial \tau_{zy}}{\partial z} = 0$$

$$\frac{\partial \sigma_z}{\partial z} + \frac{\partial \tau_{yz}}{\partial y} + \frac{\partial \tau_{xz}}{\partial x} = \gamma_0 \qquad (1)$$

be satisfied at each point of the subsoil; from the moment condition it follows that

$$\tau_{xy} = \tau_{yx}, \quad \tau_{xz} = \tau_{zx}, \quad \tau_{yz} = \tau_{zy} \qquad (2)$$

In the above equations γ_0 is the volume weight of the subsoil, σ_i and τ_{ij} are the components of the stress tensor ($i, j = x, y, z$; $i \neq j$, the pressure being positive if acting according to Fig. 2). The importance of the relations (1) and (2) for geostatic stresses was analysed by Howard (1966).

1.2.1 *Vertical geostatic stress*

For σ_z at a depth z beneath the surface of a halfspace bounded horizontally by the (x, y) plane, the third of eqns. (1) implies – taking γ_0 = const. for the sake of simplicity – that

$$\sigma_z = \gamma_0 z - \int_0^z \frac{\partial \tau_{yz}}{\partial y} \mathrm{d}z - \int_0^z \frac{\partial \tau_{xz}}{\partial x} \mathrm{d}z \qquad (3)$$

The following three cases can generally occur:

a) $$\sigma_y = \sigma_x \qquad (4)$$

This condition states that any two horizontal, mutually perpendicular normal stresses are equal to one another. In such a case, the geostatic state of stress can geometrically be represented by a vertical stress ellipsoid of revolution, and all the shear stresses τ_{ij} in eqns. (1) and (2) are zero. The normal stresses are the principal stresses. Eqn. (3) gives

$$\sigma_z = \gamma_0 z \qquad (5)$$

This case is quite usual for sediments whose state of stress is only produced by their own weight and the surface of the subsoil is horizontal (cf. Kézdi, 1962).

b) $$\int_0^z \frac{\partial \tau_{yz}}{\partial y} \mathrm{d}z = \int_0^z \frac{\partial \tau_{xz}}{\partial x} \mathrm{d}z = 0 \qquad (6)$$

The vertical geostatic stress is again defined by eqn. (5). According to Howard (1966), eqn. (6) can approximately be satisfied whenever the originally non-zero shear stress gradients relax to zero. The writer regards this possibility as very unlikely. In many samples taken from various subsoils, creep strains relax comparatively rapidly under a shear stress in the primary phase of creep, and the samples become capable of transmitting shear without further deformations. This can be explained by the fact that in the first approximation the majority of subsoils behave like Bingham's materials. Subsoils whose Bingham's threshold stress (i.e. a stress after the exceeding of which the material deforms at a constant rate) is low or zero and which, therefore, satisfy eqn. (6) most accurately of all, are of no importance as foundation soils (except in very special cases when no other construction sites are available).

It may be more to the point to assume that eqn. (6) applies to cases when – during a long period of time – a process (for example, tectonic) has probably occurred which, being opposite to the process that had produced the shear stress gradients, caused the latter to become extinct. Moreover, whenever the shear stress gradients were induced by the heterogeneity of the subsoil, their effect can be confined locally.

c)
$$\int_0^z \frac{\partial \tau_{yz}}{\partial y} \, dz = - \int_0^z \frac{\partial \tau_{xz}}{\partial x} \, dz \tag{7}$$

According to Howard (1966), the fulfilment of this condition should be regarded as highly improbable.

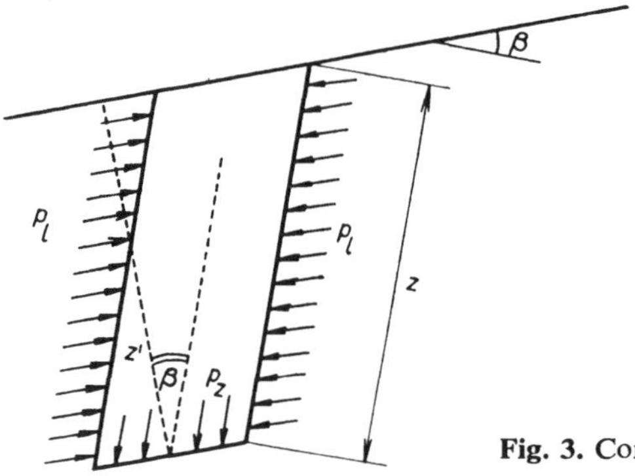

Fig. 3. Conjugate stresses p_z and p_l in an infinite slope.

A case more generally encountered than a subsoil with a horizontal surface, is a subsoil with an inclined surface, i.e. a slope. In the simplest instances, the slope can be considered to be infinitely long. As Rankine (1857, p. 20) already stated, the vertical stress p_z and the stresses parallel to the slope, p_l, are conjugate. (Fig. 3). An analysis of this state of stress can conveniently be made with the help of Mohr's circle (Taylor, 1948; Kézdi, 1962, and others). Since the stresses p_l must be of equal magnitude for reasons of symmetry, the vertical stress p_z (Fig. 3) is defined by

$$p_z = \gamma_0 z \cos \beta \tag{8}$$

where β is the slope inclination. If $z \cos \beta = z'$, then $p_z = \gamma_0 z'$, and this is in formal agreement with eqn. (5).

In Mohr's circle (Fig. 4), $\overline{OA} = p_z$, P is the pole and $\overline{OB} = \overline{OP}$ is the stress p_l in the vertical plane. Since an infinite number of circles can be drawn through the point A, the state of stress is indeterminate unless the equilibrium conditions are supplemented by additional conditions (for example, the condition of limit equilibrium or physical relations between stresses and strains, the simplest of which is Hooke's law).

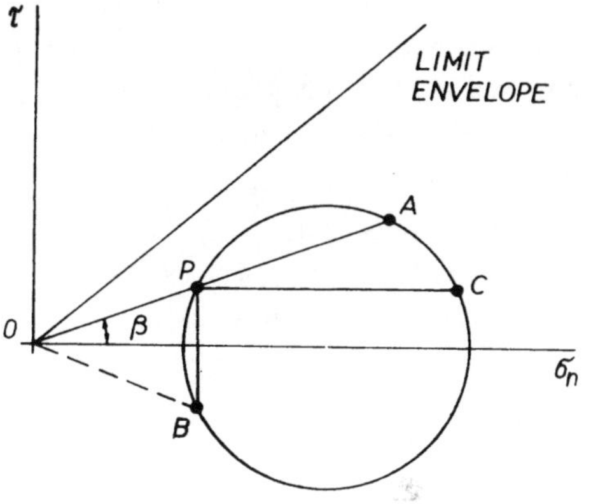

Fig. 4. A possible state of stress in a slope in Mohr's circle representation.

However, in no case (except when $\beta = 0$), is the stress $\overline{OC}$ on the horizontal plane $\overline{PC}$ a principal stress because shear stresses are acting in this plane in the same way as in the vertical planes. The shear stress gradients of an infinite slope are, however, zero (because of symmetry); hence the geostatic state of stress of such a slope is analogous to the case b) of a subsoil with a horizontal surface.

In a slope of finite length, the vertical planes near the foot of the slope (in the direction along the slope) gradually change to planes of principal stresses*) and the shear stresses vanish in them as a result. This is the way in which non-zero shear stress gradients generally arise in locations where the surface of the subsoil is vertically uneven.

The writer proposes to define the regions in which

$$\int_0^z \frac{\partial \tau_{yz}}{\partial y} \mathrm{d}z + \int_0^z \frac{\partial \tau_{xz}}{\partial x} \mathrm{d}z \neq 0 \tag{9}$$

for reasons of gravity (i.e. in consequence of the morphology of the subsoil surface), as regions of gravitational tectonic. Then a change in the regular geostatic state of stress characterized by eqn. (5) or eqn.(8), brought about by the effect of shear gradients, can be regarded as a change produced by gravitational tectonic. Such

*) Assuming the validity of eqn. (4).

a change is especially typical for denudation and accumulation processes (valley erosion, surcharge by covers of volcanic rocks, etc.).

As the foregoing analysis implies, the vertical stress σ_z is equal to the weight of the overburden only if the conditions (4) and (6) (and exceptionally, also the condition (7)) are satisfied. This applies first and foremost to sediments not disturbed by tectonic effects (at least, not after they have acquired the character of a Bingham's material), to homogeneous subsoils, and – with some reserve – to geologically old subsoils without recent tectonic. Deviations whose magnitude can often be ignored from the engineer's standpoint (for example, in view of the inaccuracies associated with the determination of the variable volume weight of the subsoil or the level of groundwater surface) arise in the neighbourhood of the bottoms of slopes or, generally, in locations where the surface of the subsoil is vertically uneven.

Another reason why only eqn. (5) should be considered as applicable to grounds formed of soils, is the fact that the measurements of in situ stresses are very delicate because the soil structure is sensitive to disturbances of any kind, changes in the state of stress, stress hysteresis, etc.

1.2.2 Horizontal geostatic stress

When the vertical geostatic stress is defined by eqn. (5), the state of stress in the subsoil is indeterminate, just as in the case of slopes.

In a very simple case when σ_z and $\sigma_x = \sigma_y$ are the principal stresses and all τ in eqns. (1) and (2) are equal to zero, eqns. (1) give

$$\sigma_x = \sigma_x(z) + \text{const.} \tag{10}$$

$$\sigma_z = \gamma_0 z \quad (\text{for } z \to 0,\ \sigma_z \to 0) \tag{5}$$

To determine eqn. (10) one must know either the constitutive equations*) (if the problem is formulated as statically indeterminate), or the condition of failure (if the problem is formulated as statically determinate). This uncertainty in the geostatic state of stress can clearly be demonstrated in Mohr's plane.

Mohr's circle is determined by a single point A (Fig. 4) which, for $\beta = 0$, lies on the axis σ_n. The only restrictive condition applicable to both cases, i.e. to both a horizontal and an inclined surface of the subsoil, is that the stress circle must not intersect the limit envelope. So long as it touches this envelope, eqn. (10) can be determined by a statically determinate solution.

For a slope with an inclination β and a cohesionless soil, the limit condition yields

*) Constitutive equations represent the relations between stress and strain, or possibly, time, which define the physico-mechanical behaviour of materials.

a pair of extreme stresses, $(p_l)_{min}$ and $(p_l)_{max}$, which are related to the vertical stress by means of the ratio (Taylor, 1948)

$$\frac{(p_l)_{min}}{p_z} = \frac{p_z}{(p_l)_{max}} = \frac{\cos\beta - \sqrt{(\cos^2\beta - \cos^2\Phi_f')}}{\cos\beta + \sqrt{(\cos^2\beta - \cos^2\Phi_f')}} \tag{11}$$

where Φ_f' is the peak (effective) angle of internal friction. The actual p_l then lies in the interval

$$(p_l)_{min} \leqq (p_l) \leqq (p_l)_{max}$$

For $\beta = 0$, i.e. for a horizontally bounded subsoil, eqn. (11) gives

$$\frac{(\sigma_x)_{min}}{\sigma_z} = \frac{\sigma_z}{(\sigma_x)_{max}} = \frac{1 - \sin\Phi_f'}{1 + \sin\Phi_f'} \tag{12}$$

Another possibility is to determine eqn. (10) by a statically indeterminate solution using the constitutive equations of the subsoil. Very simple constitutive equations are based on Hooke's law for isotropic materials, and are of the form (according to the foregoing, σ are the principal stresses)

$$\varepsilon_x = \frac{1}{E}[\sigma_x - \mu(\sigma_y + \sigma_z)]$$

$$\varepsilon_y = \frac{1}{E}[\sigma_y - \mu(\sigma_x + \sigma_z)]$$

$$\varepsilon_z = \frac{1}{E}[\sigma_z - \mu(\sigma_x + \sigma_y)] \tag{13}$$

where ε_x, ε_y, ε_z are the components of the strain tensor in the directions of the orthogonal axes (the principal strains), E is Young's modulus, and μ – Poisson's ratio. If $\varepsilon_x = \varepsilon_y = 0$ and $0 \leqq \mu \leqq 0.5$, then $\sigma_x = \sigma_y$ and

$$\sigma_x = \frac{\mu}{1 - \mu}\sigma_z \tag{14}$$

As will be shown in Section 1.2.3, more complicated constitutive equations can also be used, for example, for a cross-anisotropic halfspace with a varying direction of the axis of isotropy.

The magnitude of σ_x is usually given in the form

$$\sigma_x = K_0\sigma_z \tag{15}$$

where K_0 is the coefficient of lateral pressure (or, less exactly, the coefficient of pressure at rest). Then $\sigma_x(z) = K_0 \sigma_z$ in eqn. (10), and the respective constant is zero. By eqn. (12) it can be

$$\frac{1 - \sin \Phi'_f}{1 + \sin \Phi'_f} \leqq K_0 \leqq \frac{1 + \sin \Phi'_f}{1 - \sin \Phi'_f} \tag{16}$$

This possibility is generally admitted (Lambe and Whitman, 1969; Morgenstern and Eisenstein, 1970). For cohesive soils, Florin (1959, p. 95) formulates it as follows:

$$\sigma_z \operatorname{tg}^2\left(\frac{\pi}{4} - \frac{\Phi'_f}{2}\right) - 2c' \operatorname{tg}\left(\frac{\pi}{4} - \frac{\Phi'_f}{2}\right) \leqq \begin{matrix}\sigma_x \\ \text{or} \\ \sigma_y\end{matrix} \leqq \sigma_z \operatorname{tg}^2\left(\frac{\pi}{4} + \frac{\Phi'_f}{2}\right) + 2c' \operatorname{tg}\left(\frac{\pi}{4} + \frac{\Phi'_f}{2}\right) \tag{17}$$

Although, in keeping with eqn. (14),

$$0 \leqq K_0 \leqq 1 \tag{18}$$

for $0 \leqq \mu \leqq 0.5$, K_0 actually varies within much broader limits.

A great variability of the coefficient K_0 observed in nature, can be explained from a physical standpoint by the fact that in geological time the subsoil experiences a complicated deformation history connected with complex loading and unloading cycles. These cycles leave their traces in its structure (the subsoil has a "memory"), for example, in the form of residual (or better yet, internal) stresses (cf. Voight, 1966; Feda, 1971b, and others; Hruban, 1945, p. 22, uses the term "secondary stresses"), and the state of stress can eventually reach the limit state. Besides the vertical stress, the horizontal stress can also undergo a change. An interesting idea is put forward by Price (1969) who supposes that a progressive lateral compression generally takes place in a sedimentation basin during the accumulation of sediments. However, he considers it unlikely that the horizontal stresses could exceed the vertical stresses in the process.

The present-day ideas concerning the magnitude of the coefficient K_0 are based on theoretical considerations of Jáky (1944), which are quoted extensively by Kézdi (1962). In place of the original theoretical relation, use is made of Jáky's simplified expression

$$K_0 = 1 - \sin \Phi'_f \tag{19}$$

(where Φ'_f is the effective peak angle of internal friction), which is in better agreement with experimental results (Bishop, 1958; Kézdi, 1962, 1974; Hendron, 1963; Wroth, 1972, and others). Jáky derives the magnitude of K_0 from the state of stress along the axis of a triangular embankment, with the slopes inclined at an angle of Φ'_f and an empirical distribution of the shear stress in its central section. It is hard to concur with the idea that such assumptions are acceptable for deriving the state of stress in horizontally bounded subsoils (Tschebotarioff, 1951). Jáky's relation is, therefore, of an empirical character.

Eqn. (19) is employed for normally consolidated cohesionless soils. Results of experiments made with normally consolidated cohesive soils (with an effective cohesion $c' = 0$) are better described by the equation (Brooker and Ireland, 1965; Lambe, 1967, and others)

$$K_0 = 0.95 - \sin \Phi'_f \tag{20}$$

Comparing eqns. (19) and (20) leads to the more general form (Feda, 1971a)

$$K_0 = a - \sin \Phi'_f \tag{21}$$

where a depends on the soil structure, the loading path, etc. Using the above relations, one can determine K_0 to an accuracy of about ± 15 to $\pm 20\%$.

Relations (19) to (21) are sometimes understood to mean that a considerable part of shear strength is mobilized during the compression of the subsoil (Kézdi, 1962), as theoretically happens in the case of regular arrays of spheres (Hendron, 1963; Raju, 1967). However, a real statistical medium behaves differently. It is easy to demonstrate a correlation between the angle of internal friction and porosity, and between porosity and compressibility, and define K_0 as a function of compressibility (such a relation was theoretically derived by Roscoe and Burland, 1968, for normally consolidated clays). Since they reflect the effect of the structure in the broadest sense of the word, all the basic mechanical parameters are mutually related. Expressing one parameter in terms of another can, however, be questionable because some parameters involve different stress paths and stress levels, and this fact is apt to lead to different changes in the soil structure. For this reason Bjerrum and Andersen (1972) regard the coefficient K_0 as an independent basic parameter.

Hence, for normally consolidated soils one can use the equation

$$\sigma_x = (a - \sin \Phi_f)\, \sigma_z \tag{22}$$

where $a = 1$ for cohesionless, and $a = 0.95$ for cohesive soils.

So far as their geostatic state of stress is concerned, overconsolidated subsoils can be considered in the same way as normally consolidated subsoils, i.e. the axis of the ellipsoid of their in situ stresses is assumed to be vertical. In some cases, such an assumption may be disputable, for the process of consolidation — in the same way as for tectonic processes (Section 1.3) — is apt to bring about a marked change in the orientation of the principal axes of stresses. In overconsolidated subsoils, the magnitude of the coefficient K_0 depends on the overconsolidation ratio (OCR), which is the ratio between the maximum vertical effective stress in the geological history and the present vertical effective stress ($OCR = 1$ for normally consolidated, and $OCR > 1$ for overconsolidated subsoils).

The effect of unloading on an increase of K_0 can be determined in laboratory (Fig. 5). It is clearly seen that overconsolidation affects the magnitude of K_0 more noticeably than do other factors (for example, the relative density of sand causes K_0

to vary approximately between 0.3 and 0.5). An idea of how K_0 increases in situ can be gained by referring to Fig. 6.

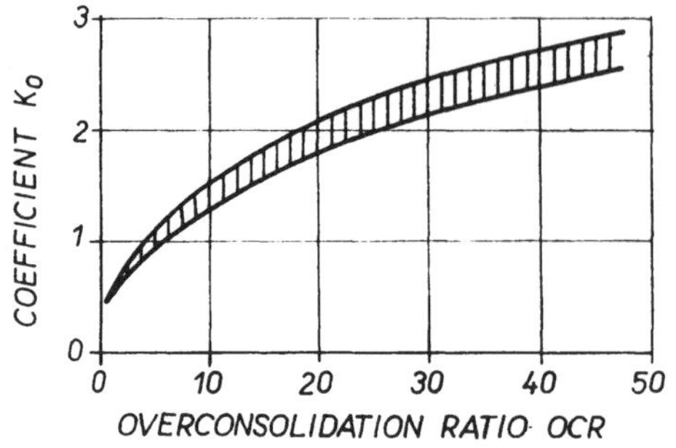

Fig. 5. Relation between the coefficient K_0 and the overconsolidation ratio OCR of a Pennsylvania sand (Hendron, 1963).

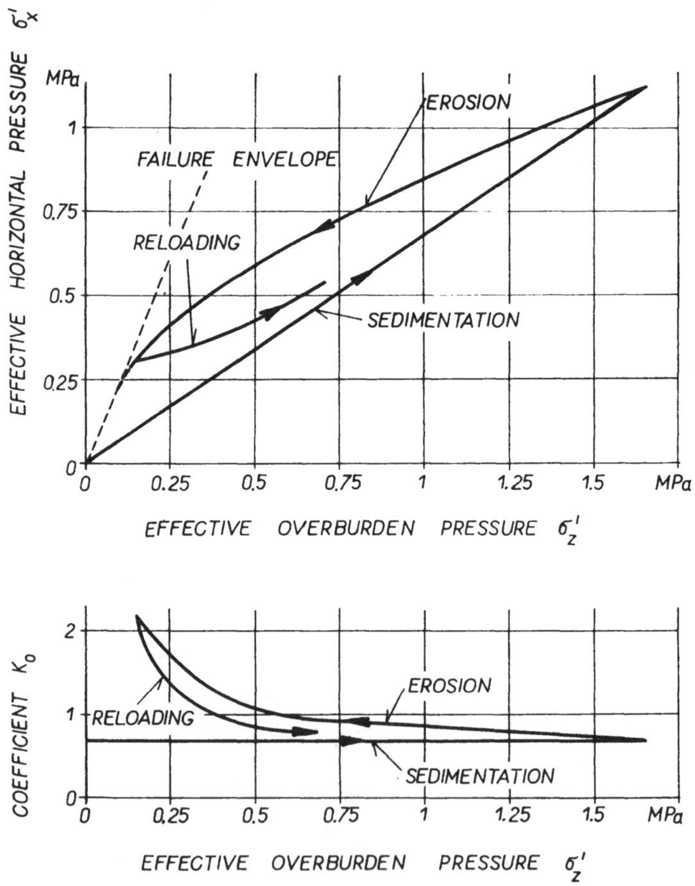

Fig. 6. Relation between vertical and horizontal in situ stresses during sedimentation and erosion cycles.

Clays overconsolidated by desiccation (Moretto, 1969)*) or overconsolidated subsoils at a greater depth are apt to have $K_0 \to 1$ when the in situ stress tends to be isotropic (Fig. 7).

The idea that K_0 can be larger than unity, was put forward by Samsioe in the thirties (Skempton, 1961). By reconstructing the geological history of London clay in Bradwell, Skempton (1961) showed that in this case, $K_0 > 2.5$ and that the 10–15 m

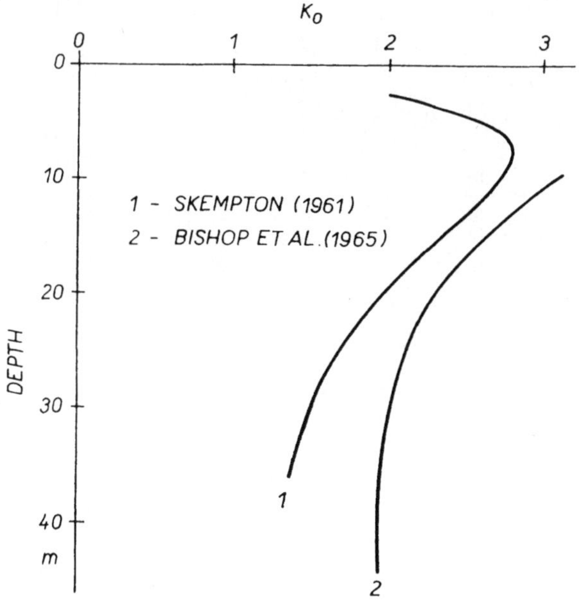

Fig. 7. Variations of the coefficient of lateral pressure K_0 with depth in overconsolidated London clay, measured in laboratory on undisturbed samples (height of eroded overburden: about 150 m — (1); about 440 m — (2)).

thick upper layer of the subsoil is at the passive limit state of stress (cf. Terzaghi, 1961). The results of Skempton and Bishop et al. (1965) are reproduced in Fig. 7. Lambe and Whitman (1969, p. 100) concede that in overconsolidation, K_0 can reach as high a value as 3. Brooker and Ireland (1965) claim that the magnitude of K_0 is not only a function of *OCR* but a function of the value of the angle of internal friction

---

*) Provided that no shrinkage cracks appear in the subsoil,

$$K_0 = \frac{p_k}{\gamma_n z + p_k} \quad \text{for} \quad z \to 0, \ K_0 \to 1$$

(p_k — the omnidirectional capillary pressure).

If shrinkage is followed by re-saturation, the lateral swelling resulting from irreversible changes in the structure might be too small to mobilize the original horizontal geostatic pressure; in such cases, K_0 is likely to be even less than that of normally consolidated soils.

as well. For $OCR > 8$, the effect of Φ'_f is not unambiguous (Fig. 8). According to Ladd (1964), K_0 becomes greater than unity for $OCR > 2.5-4.5$.

Wroth (1972) deduced an empirical relation which can be given the following form:

$$\lg OCR = (3.6 + 0.07\, I_p)\left(\frac{1 - K_0^n}{1 + 2K_0^n} + \frac{K_0 - 1}{1 + 2K_0}\right) \tag{23}$$

where I_p is the plasticity index (in %), K_0 and K_0^n is the coefficient K_0 of an overconsolidated and a normally consolidated soil, respectively (cf. eqns. (19), (21)). For sands, eqn. (23) yields values which are larger by far than those shown in Fig. 5; for these materials it only has a physical meaning when $OCR < 20$.

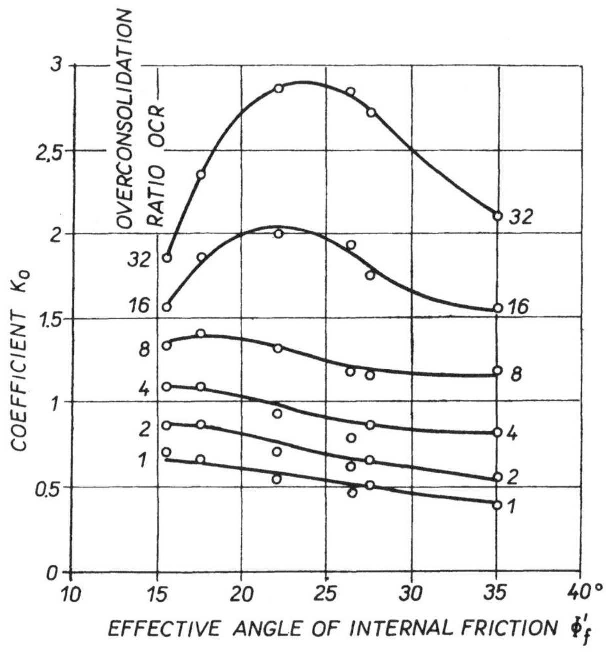

Fig. 8. Relation between the coefficient K_0, the overconsolidation ratio OCR and the effective angle of internal friction (Brooker and Ireland, 1965).

For overconsolidated clays, Sherif (Tschebotarioff, 1973) proposed a relation which is somewhat simpler than eqn. (23): K_0 is a function of OCR and of two parameters which depend on the liquid limit of the clay in question.

The effect of the value of OCR on K_0 is expressed yet more simply by Caquot and Kérisel (1966, p. 112) using the empirical relation $K_0 = 0.5\sqrt{(OCR)}$. This relation is more frequently employed in the form

$$K_0 = K_0^n(OCR)^b \tag{24}$$

where $b = 0.39$ to 0.58 (for clays and sands as well; for sands, b is independent of the relative density); in undisturbed samples it increases, in remoulded samples it decreases with increasing I_p (Belotti et al., 1975). Eqn. (24) is fittingly expressed by Fig. 5.

To complete the discussion, it may be well to mention the fact that the magnitude of K_0 is affected, for example, by compaction of loose soils in the back of a supporting wall, in a similar way as it is by *OCR* (Broms, 1971 – K_0 in situ ranged between 0.9 and 1.5, in laboratory between 0.45 and 2). A similar effect was also observed during pile driving, even for piles of constant diameter along their whole length (Bassett, 1970, estimates that K_0 increases to 1 to 3.5 in the course of the process).

By applying the finite element method, Malina (1969) succeeded in numerically determining the effect of overconsolidation on the magnitude of K_0. He reports that in the vicinity of the surface, the state of stress is close to passive; this finding is in agreement with an analysis published by Skempton (1961).

1.2.3 The effect of anisotropy and structure on K_0

Similarly as in the case of eqn. (14) one can deduce K_0 for an anisotropic material. This can best be done in the simplest case of cross (axi-symmetrical) anisotropy. In this case, five parameters are usually needed to describe the mechanical behaviour of a subsoil. If the principal axis of the stress tensor as well as the axis of isotropy are vertical – as they are assumed to be in the case of K_0 – the following four parameters are sufficient for the purpose: Young's modulus in the vertical (E_v) and in the horizontal (E_h) direction, and three Poisson's ratios (μ_1 – the effect of horizontal stress on horizontal strain, or generally, on strain in the plane of isotropy, μ_2 – the effect of horizontal stress on vertical strain, μ_3 – the effect of vertical stress – in the axis of isotropy – on horizontal strain – in the plane of isotropy; $\mu_2 = E_h\mu_3/E_v$).

In place of eqns. (13) one obtains for the principal stresses and strains the equations

$$\varepsilon_x = \frac{\sigma_x}{E_h} - \frac{\mu_1}{E_h}\sigma_y - \frac{\mu_3}{E_v}\sigma_z$$

$$\varepsilon_y = \frac{\sigma_y}{E_h} - \frac{\mu_1}{E_h}\sigma_x - \frac{\mu_3}{E_v}\sigma_z$$

$$\varepsilon_z = \frac{\sigma_z}{E_v} - \frac{\mu_2}{E_h}\sigma_x - \frac{\mu_2}{E_h}\sigma_y \qquad (25)$$

For the condition set forth for K_0 (i.e. $\varepsilon_x = \varepsilon_y = 0$) the above equations give $\sigma_x = \sigma_y$, and

$$K_0 = \frac{E_h}{E_v}\frac{\mu_3}{1 - \mu_1} \qquad (26)$$

Eqn. (26) applies to the case when the axis of isotropy is vertical. For a horizontal axis of isotropy, one obtains, analogously, (the stresses σ_x, σ_y, σ_z are principal but $\sigma_x \neq \sigma_y$) that

$$K_0 = \frac{\sigma_x}{\sigma_z} = \frac{\mu_3(1 + \mu_1)}{1 - \frac{E_v}{E_h}\mu_3^2} \tag{27}$$

and

$$K_0 = \frac{\sigma_y}{\sigma_z} = \frac{\mu_1 + \frac{E_v}{E_h}\mu_2^2}{1 - \frac{E_v}{E_h}\mu_2^2} \tag{28}$$

In the case of isotropy ($E_v = E_h$, $\mu_1 = \mu_3$), eqns. (26), (27) and (28) take the form of eqn. (18). They were derived by Uriel (1973) and Doležalová (1974).*)

If the axis of isotropy is oblique, the principal axis of the stress tensor is no longer vertical, and the fifth strain parameter G (shear modulus in planes normal to the plane of isotropy) comes into play. A solution of the case of the axis of isotropy making an angle of 45° with vertical, was obtained by Doležalová (1974) using the finite element method. Her results are reproduced in Fig. 9.

The effect of anisotropy is most marked for a vertical axis of isotropy (0° in Fig. 9). But even in this case, K_0, within a realistic range of the ratio between Young's moduli, is substantially smaller than its in situ value (Fig. 7). Hence the large values of K_0 should be accounted for by the inelastic behaviour rather than by the anisotropy of the subsoil.

The magnitude of K_0 can be determined by a structural or a phenomenological procedure. The procedure so far employed was phenomenological, i.e. the soil was regarded as a quasi-continuum, and the concepts of continuous stresses and strains were used throughout. The structural procedure is based on statical and deformation relations existing between the various particles of an elementary volume of the soil. The resulting relationships are then generalized to cover the entire volume of the sample; this is done by changing from a discrete distribution of the contact forces to a continuously defined stress.

The structural procedure was applied to the determination of the coefficient K_0 of cohesionless soils by Hendron (1963) and particularly, by Raju (1967) from whose solution Hendron's relations can be deduced as a special case. On the basis of the results obtained by a number of authors (Mindlin and Deresiewicz, 1953; Duffy,

---

*) Eqns. (26) and (28) can also be expressed in terms of Poisson's ratios alone, viz.

$$K_0 = \mu_2/(1 - \mu_1) \quad \text{and} \quad K_0 = (\mu_1 + \mu_2\mu_3)/(1 - \mu_2\mu_3)$$

1959; Duffy and Mindlin, 1957; Thurston and Deresiewicz, 1959), they model a cohesionless medium by a regular set of equal diameter spheres (monodispersive set) arranged in a cubical and a hexagonal array. The most important of the relations thus obtained are those relating to the two systems in their most compact arrangement (a porosity of 25.95%)*) because they were verified experimentally by Raju (1967).

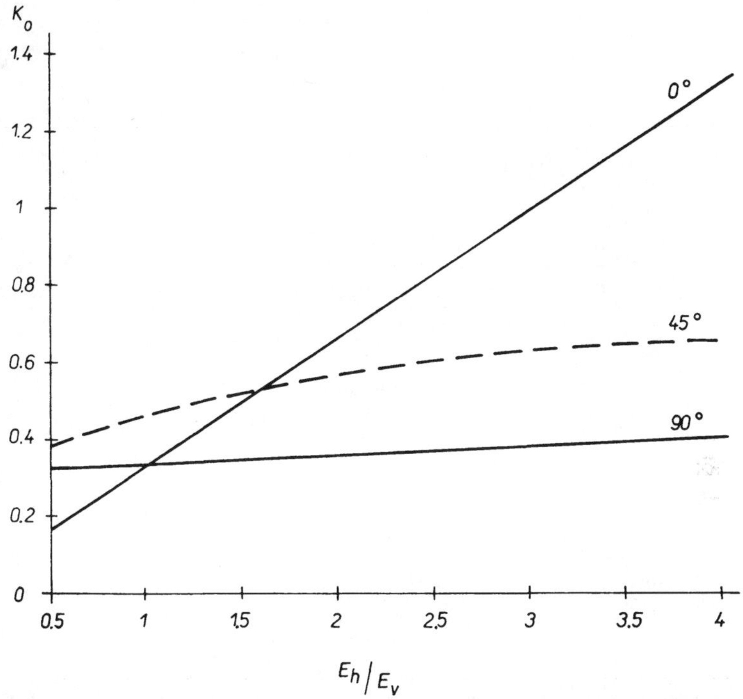

Fig. 9. Effect of cross anisotropy on the magnitude of K_0 — Doležalová (1974). The angle between the vertical and the axis of isotropy is marked. The following parameters were used in the calculations: $\mu_1 = \mu_3 = 0.25$; $G = 102.2$ MPa = const.; $\gamma_0 = 24.5$ kN/m^3; ratio $E_h/E_v = 0.5$, 2 and 4 ($E_v = 490$, 245 and 122.5 MPa).

From the conditions of equilibrium of the contact forces, the conditions of compatibility of the displacements (a mathematically formulated condition of the constancy of contacts in the deformation process), from the equation for normal and shear compliance of the contacts and from the boundary conditions, one can derive the equations for uniaxial and hydrostatic compression and the coefficient K_0 under a gradually increasing and decreasing load. As Hendron (1963) and Raju (1967)

*) In such an arrangement, the number of contacts of each sphere with the lower, medium and upper layer is the same, [4, 4, 4] for the cubical array, and [3, 6, 3] for the hexagonal array. The two arrays do not differ except by orientation: the former can be obtained from the latter by rotation.

established, for a model medium, the coefficient K_0 is a function of the angle of intergranular friction, Φ_μ and of the angle α which the normal to the contact plane of the spheres makes with the horizontal plane, viz.

$$K_0 = \frac{1}{2\,\mathrm{tg}\,\alpha}\,\frac{1 - \mathrm{tg}\,\Phi_\mu\,\mathrm{tg}\,\alpha}{\mathrm{tg}\,\alpha + \mathrm{tg}\,\Phi_\mu} \tag{29}$$

($\mathrm{tg}\,\alpha = 1$ and $\sqrt{2}$ for the most compact cubic and hexagonal array, respectively). According to Wittke, the following relation can be derived for the limit ratio between the axial and the radial stress $(\sigma'_a/\sigma'_r)_f$ of the cubic set, which defines the peak angle of internal friction Φ'_f (Raju, 1967, p. 75):

$$\left(\frac{\sigma'_a}{\sigma'_r}\right)_f = 2\,\mathrm{tg}\,\alpha\,\frac{\mathrm{tg}\,\alpha + \mathrm{tg}\,\Phi_\mu\,\dfrac{1}{\sqrt{(1+\sin^2\alpha)}}}{1 - \mathrm{tg}\,\Phi_\mu\,\mathrm{tg}\,\alpha\,\dfrac{2}{\sqrt{(1+\sin^2\alpha)}}} \tag{30}$$

A special case of eqn. (30) for $\alpha = 45°$ is mentioned by Thurston and Deresiewicz (1959). By combining eqns. (29) and (30) one obtains

$$K_0 = \frac{1}{2}\,\frac{1 - \frac{1}{4}\sqrt{(1+\sin^2\alpha)}\,[(1 + 2\,\mathrm{tg}^2\,\alpha)\sin\Phi'_f - (2\,\mathrm{tg}^2\,\alpha - 1)]}{\mathrm{tg}^2\,\alpha + \frac{1}{4}\sqrt{(1+\sin^2\alpha)}\,[(1 + 2\,\mathrm{tg}^2\,\alpha)\sin\Phi'_f - (2\,\mathrm{tg}^2\,\alpha - 1)]} \tag{31}$$

Hendron's relation for K_0 (cf. Lambe and Whitman, 1969, p. 128)

$$K_0 = \frac{1}{2}\,\frac{1 + \dfrac{\sqrt{6}}{8} - \dfrac{3\sqrt{6}}{8}\sin\Phi'_f}{1 - \dfrac{\sqrt{6}}{8} + \dfrac{3\sqrt{6}}{8}\sin\Phi'_f} \tag{32}$$

is a special case of eqn. (31) for $\alpha = 45°$.

Numerous experiments have yielded information concerning the close correlation between K_0 and Φ'_f, and between Φ'_f and the initial specimen porosity n_0, which implies the correlation between K_0 and n_0. Eqn. (31) also relates K_0 to Φ'_f for various values of the angle α for which the model cubic set is stable. Since the relation between n_0 and α of the model sets is neither unique (for the same α, n_0 of the cubic set is different from that of the hexagonal set) nor monotonic (cf., for example, Raju, 1967, pp. 17 and 67), the relation between K_0 and n_0 cannot be derived from eqn. (31). Thus, for example, for $n_0 \doteq 26\%$ and $\mathrm{tg}\,\Phi_\mu = 0.1$, $K_0 = 0.409$ for the most compact cubic array, and $K_0 = 0.201$ for the most compact hexagonal array.

For $\alpha \doteq$ const, the porosity of a monodispersive deterministic set of spheres is constant. Hence the dependence of K_0 on Φ'_f defined by eqn. (31) or eqn. (32), is exclusively a result of the assumed variability of the angle of intergranular friction

Φ_μ. In a real cohesionless medium, a given sample has an average Φ_μ = const., and the relation between K_0 and Φ_f' is a result of the different density of the sample. In a regular set of spheres, the density is assumed to be constant, and $\Phi_\mu \neq$ const. Hence a direct application of the results of analyses of model media to real cohesionless materials is not satisfactory on physical grounds, and the circumstance that eqn. (32) was confirmed by the experimental values of K_0 of a uniform sand with spherical grains, reported by Hendron (1963), should be regarded as a mere accidental concurrence.

The reason why a deterministic model of spheres is physically inadequate for describing the behaviour of real cohesionless media can be traced back to the assumption made in the interest of the solvability of the problem, i.e. that all the contacts of the spheres are fixed and that their number is constant throughout the deformation process. Hence a deterministic set does not deform by mutual sliding of grains (intergranular displacements), which is the main source of strains in a real cohesionless medium (provided that the grains are not crushed), but merely by intrinsic strains of the individual spheres in contact (intragranular strains). This is the reason why an axial strain, ε_a, of a deterministic medium under an axial load, σ_a, is of two orders of magnitude (ε_a is equal to hundredths to thousandths of percent) smaller than that of a real cohesionless medium, for example, sand (cf. Thurston and Deresiewicz, 1959 – Fig. 4; Raju, 1967 – Fig. 23, and others).

Thus the essential quality of the physical differences between a model and a real medium lies in the different deformation mechanism of the two media – exclusively intragranular in the former case, and exclusively intergranular in the latter. This fact is changed in no way by the introduction of partial improvements of the deterministic set by a bi-dispersive model in which the number of contacts increases with the load (Schlosser, 1974), or by a model medium composed of regions formed of deterministic sets in a cubic and a hexagonal array.

The fact that some of its results are in qualitative agreement with the actual behaviour of cohesionless materials, is sometimes mentioned in support of the theory of deterministic sets. Thus, for example, one can deduce for a model an exponential relation between ε_a and σ_a under uniaxial (and similarly, under hydrostatic) compression which corresponds to the character of the compressibility of sand. The same relation is, however, implied by Hertz' formulae defining a change of the distance between the centres of two spheres pressed together, which is certainly a physically unacceptable model of a cohesionless medium. In fact, the theory of deterministic models is an essentially insignificant improvement of these formulae.

The theory of deterministic sets applies accurately enough to media formed precisely like the model. If the spheres are of exactly the same diameters so that the structure of the set is regular, experiments corroborate the theoretical character and magnitude of compression. Raju (1967) even succeeded in proving an identity between the experimental and the theoretical magnitude of K_0 for the most compact cubic and hexagonal array of steel balls under increasing loading when, as is well known,

K_0 = const.*), and in obtaining a reasonable agreement with experiments during unloading, thus confirming the familiar dependence of K_0 on the overconsolidation ratio (cf. Fig. 5, eqns. (23) and (24)).

Raju's (1967) clarification of the variability of K_0 during uniaxial unloading goes theoretically much deeper than Hendron's (1963). Hendron adapted the same procedure for unloading as for loading except that he replaced $+\text{tg}\,\Phi_\mu$ by $-\text{tg}\,\Phi_\mu$, Raju (1967), on the other hand, makes use of the complete relation for shear compliance under a monotonic and simultaneous decrease of the shear and normal components of the contact force, evolved by Mindlin and Deresiewicz (1953). The dependence of the model medium on the stress path is induced by the elastic-plastic nature of shear compliance. Since the normal compliance is only elastic, it is independent of the stress path. Hence, for example, hydrostatic strains in a model medium are purely elastic.

It was revealed in experiments with model media that the value of Φ_μ changes from one particle to another (cf. Raju, 1967, pp. 87, 88) so that the measurement of Φ_μ calls for a phenomenological procedure, i.e. tests of the whole sample, which yields an average and statistically significant value of Φ_μ (Procter and Barton, 1974).

As an analysis of the structural method for determining the coefficient K_0 clearly shows, the method is not only unacceptable on physical grounds but the decisive parameter Φ_μ must be established phenomenologically.

Analyses so far performed relate to structural models of a cohesionless medium, i.e. a medium composed of grains (structural units) which are smaller by far than the size of a sample. One can also imagine a physically similar discontinuous medium, i.e. a particulate material whose structural units are much larger than the size of a sample. As an example of such a material, consider a rock mass penetrated through by fissures, in which the structural units are formed by blocks of the rock. A mathematical model of this medium, consisting of intergranular (joint) elements and intragranular elements (blocks) was constructed by means of the finite element method (Goodman et al., 1968) – cf. Section 3.8.

An analysis of the state of stresses and strains in such a medium is essentially also a structural analysis. The strains are produced in part by the displacements of the rock blocks – intergranularly, in part by deformation or possible crushing of the blocks – intragranularly. The stresses are also of two kinds, and the geostatic stress which has so far been considered uniform throughout, should be divided into the geostatic stress inside the joints and the geostatic stress inside the blocks – in logical consequence of the structural conception. To the best of the writer's knowledge, such a solution has not been presented so far, although due regard was already paid to the scale effect on the magnitude of the geostatic stress in a rock mass (Gaziev, 1973, p. 88).

Since a solution of the geostatic state of stress in soils obtained by the structural

*) In loading, the medium is normally consolidated, i.e. $OCR = 1 =$ const.

procedure is not successful, and the phenomenological relations (19) to (24) are only of an empirical nature, an adequate theoretical solution is, in fact, not available. But, as is well known, the coefficient K_0 of soils depends on the soil structure (the effect of n_0, I_p, Φ_f'), the stress path (the effect of OCR), etc.; hence only a phenomenological relation that could be interpreted structurally, ought to be theoretically acceptable. In contrast to the empirical relations, it should make it possible to extrapolate K_0 rather than just to summarize the known experimental results.

These requirements, in part at least, are met by the writer's phenomenological procedure based on the theory of plastic potential because a plastic strain is the most significant of all components of the total strain of soils. The relation obtained by the writer is

$$K_0 = \frac{1 - \sin \Phi_r}{\sin \Phi_r + 2 \dfrac{d\varepsilon_a^e}{d\varepsilon_a}} \tag{33}$$

where Φ_r is the residual angle of internal friction; $d\varepsilon_a^e$, $d\varepsilon_a$ is the increment of the reversible and the total axial strain, respectively. The quantity Φ_r is related to Φ_μ. Tests of eqn. (33) made for sand showed the maximum deviation from Jáky's relation (19) to be -20% for dense sand (while full agreement was obtained for loose sand); the value of the $d\varepsilon_a^e/d\varepsilon_a$ ratio was measured in an oedometer. When this ratio increases, the structure grows stronger; when it decreases, the structure shows a tendency towards failure. Hence the coefficient K_0 is directly related to the changes in structure. If K_0 = const. (in normally consolidated soils), the processes of strengthening and of weakening of the structure are in equilibrium. The same takes place under hydrostatic compression. Eqn. (33) enables some predictions to be made beyond the common experimental evidence. Thus, for example, the $d\varepsilon_a^e/d\varepsilon_a$ ratio decreases by the effect of grain crushing and time; hence K_0 will increase with growing stress (when it is high enough to cause crushing) and time. In another case one can predict the magnitude of K_0 of two cohesionless materials that differ only by the values of Young's moduli of their grains. In keeping with eqn. (33), the material with a higher Young's modulus of the grains will have a larger K_0 (the same conclusion is reached by Andrawes and El-Sohby, 1973). Eqn. (33) and the measured values of $d\varepsilon_a^e/d\varepsilon_a$ during oedometric unloading also imply the variability of K_0 during unloading (i.e. at $OCR > 1$).

1.2.4 Conclusion

As both the theoretical and the experimental results suggest, the geostatic state of stress can usually be characterized by the equations

$$\sigma_z = \gamma_0 z \tag{5}$$

and

$$\sigma_z(a - \sin \Phi_f') \leqq \sigma_x \leqq \sigma_z \, \mathrm{tg}^2 \left(\frac{\pi}{4} + \frac{\Phi_f'}{2}\right) + 2c' \, \mathrm{tg} \left(\frac{\pi}{4} + \frac{\Phi_f'}{2}\right) \tag{34}$$

($a = 1$ for cohesionless soils, $a = 0.95$ for cohesive soils).

The horizontal stresses $\sigma_x = \sigma_y$ increase from the lower limit (applicable to normally consolidated subsoils) to the upper limit (the passive limit state of stress) proportionally to the increasing overconsolidation ratio, *OCR*. The upper limit of the coefficient K_0 probably also depends on a number of additional factors (Brooker and Ireland, 1965, have shown the effect of the angle of internal friction by laboratory experiments). The magnitude of K_0 is a direct reflection of the structure and its changes during a uniaxial deformation process; it is affected by the anisotropy of the medium, etc. It is, therefore, generally acknowledged that, for the time being, a reliable a priori knowledge of in situ stresses is seldom within one's reach, except, perhaps, in the case of normally consolidated clays (Morgenstern and Eisenstein, 1970).

At present, the solution of the geostatic state of stress is predominantly done on a semi-empirical level. An exact solution calls for the knowledge of the constitutive equations, including the effect of stress and strain history. Although promising results have been obtained by means of the finite element method, this possibility is still a long way off. The structural approach to discontinuous media (Goodman et al., 1968) merits special attention.

1.3 Geodynamic stresses

Since tectonic activity induces high lateral pressures in the ground, its effect in this respect can be compared to that of overconsolidation. It can be likened to the effect of pile driving, which is mentioned by Bassett (1970). Geodynamic stresses concentrate mainly in the rigid elements of the earth crust, i.e. in rocks in which stresses relax slowly. According to a statistical analysis, $K_0 \geqq 2$ is frequently encountered in rocks (Voight, 1966), and a considerable difference is observed to exist in geostatic stresses on the two sides of a fault or a failure zone (Gaziev, 1973, pp. 86, 90). Because of their low compressibility, rocks are, however, unimportant from the point of view of settlement.

On the other hand, studies of geodynamic stresses in rocks can be used to advantage in investigations concerning more compressible subsoils for the following two reasons:

a) they indicate the qualitative effect of geodynamic stresses on the in situ state of stress whenever the quantitative effect is either small or hard to separate from the effect of overconsolidation;

b) the methods employed for analysing the in situ state of stress in rocks are also used for soils sometimes (cf. Section 1.7).

Shear stresses typical of tectonic processes lead to the applicability of eqn. (9). In this case, the principal axes of the ellipsoid of the in situ stresses are rotated, and the ellipsoid can no longer be determined on the basis of the σ_z, σ_x, σ_y stresses. The mea-

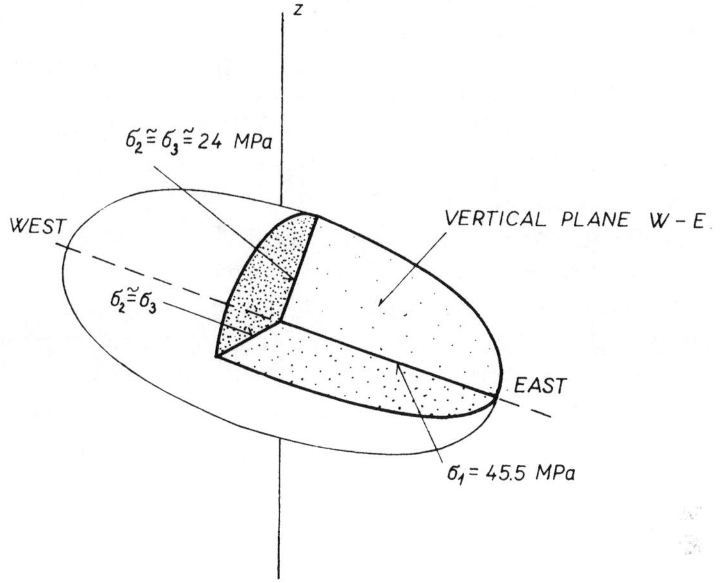

Fig. 10. Ellipsoid of in situ stresses in limestone — according to measurements made in Barberton (Obert et al., 1962).

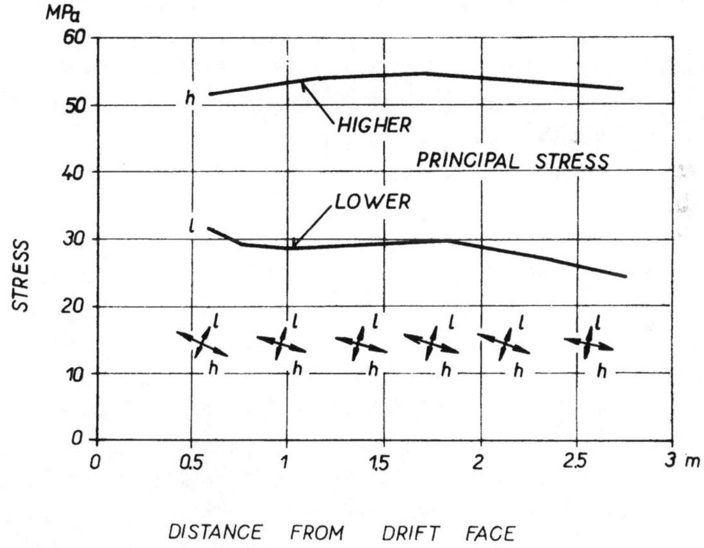

Fig. 11. Average results of stress measurements made in vertical plane with east-west orientation in Barberton (Obert et al., 1962).

surements of the in situ stresses made by Obert et al. (1962) in a limestone quarry in Barberton (and quoted by Voight, 1966) can serve as an illustrative example: the ellipsoid of the in situ stresses constructed on their basis, is shown in Fig. 10, and the inclination of the axis of the principal stress in the east-west direction is marked in Fig. 11. Similar results were published by Gaziev (1973, p. 90).

Some overconsolidated subsoils consisting, for example, of stiff to hard clays and clay slates are likely to show qualitatively similar patterns of the in situ stresses. Studies of sedimentary fills of brown coal fields point to the existence of a considerable additional horizontal stress produced by overconsolidation and tectonically conditioned vertical and horizontal movements (Rybář, 1971). Esu and Calabresi (1969) presented a hypothesis according to which a change of the orientation of the principal axes of the geostatic stresses in layers of overconsolidated pliocene clays was brought about by the circumstance that the layers were bent by the effect of compression of the underlying peat. It should, of course, be borne in mind that the principal axis of the stress tensor can be oblique because of the effect of anisotropy – for example, when the axis of isotropy of a cross-anisotropic subsoil is neither vertical nor horizontal (Section 1.2.3).

1.4 Neutral stresses

It is assumed in the discussion that follows that the in situ stresses are defined by eqns. (5) and (15). In keeping with eqn. (5), σ_z increases linearly with depth. If $\gamma_0 = \gamma(z)$ – as in the case of stratified subsoils – the simple straight line which represents the relation between σ_z and the depth is replaced by a broken line. If a continuous surface of groundwater reaches up to the surface of the subsoil, the effective stress is the factor which decides about the amount of the settlement (and about all mechanical processes which take place in the subsoil). This also holds true of the in situ stresses. If this is the case, then

$$\sigma'_z = \gamma_n z - u \tag{35}$$

where σ'_z is the effective vertical stress, u – the neutral (pore) stress, i.e. in the case being discussed, the water pressure in the pores, γ_n – the unit weight of the saturated soil. The neutral stress is usually compression (marked with the + sign). When the effect of capillary stress is also taken into account, the neutral stress in the region of capillarity is tension (marked with the – sign); in this case, the σ'_z stress is higher than the total stress (Kézdi, 1962; Lambe and Whitman, 1969, p. 380). In large foundations, this effect usually plays no special role, and its introduction in the calculations is not straightforward. When a subsoil is incompletely saturated with water, the general principle of effective stresses (Bishop, 1961) in which the pressure of air in pores also figures, should be used in place of eqn. (35), and γ_0 (the unit weight of naturally wet soil) should be substituted for γ_n.

Referring to Figs. 12 and 13, the following basic cases can be distinguished (Kögler and Scheidig, 1938; Taylor, 1948; Kézdi, 1962, etc.):

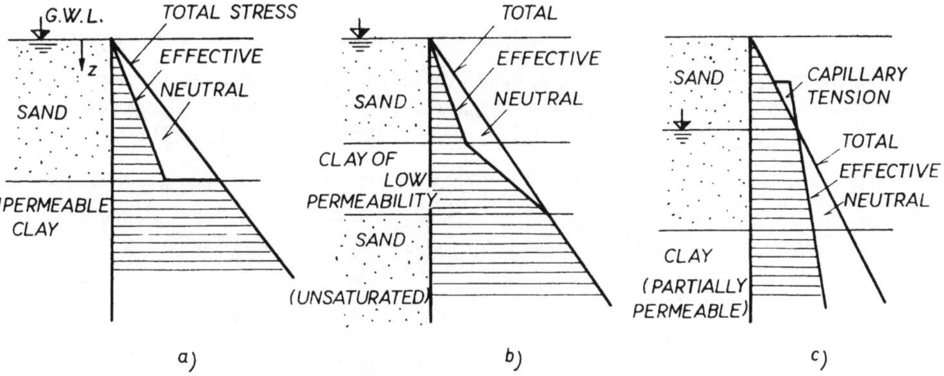

Fig. 12. Schematic diagram of vertical effective stresses in foundation soil.

Fig. 12a: The neutral stress (uplift) $\gamma_w z$ (γ_w – the specific weight of water) acts in a layer of sand; hence

$$\sigma'_z = (\gamma_n - \gamma_w)\, z = \gamma' z \tag{36}$$

The quantity $\gamma_n - \gamma_w = \gamma'$ is the unit weight of the submerged soil. If the clay is impervious, the neutral stress acting in it is $u = 0$ (for the sake of simplicity, the unit weight of sand is assumed to be equal to the unit weight of clay). A sudden step in the geostatic pressure is sometimes smoothed over by the assumption that the uplift experiences a gradual decrease to a certain depth below the surface of clay. The form of the geostatic stress in clay shown in Fig. 12a is used in explanations relating to the settlement of impervious bottoms of dam reservoirs (cf. Mencl, 1955; and recently, Mathian and Paubel, 1969). Since dense clays are of very low permeability, their neutral stress is extremely difficult to measure. This may be the reason why it is sometimes supposed that the uplift in such soils is either completely absent or, at least, reduced (Florin, 1959, p. 99). Such an opinion can be refuted on both theoretical and experimental grounds (Bažant, 1938, pp. 17 to 20).

Fig. 12b: The uplift acts in a layer of sand but groundwater seeps through a layer of low-permeability clay so that it raises its geostatic stress by the flow pressure. No continuous goundwater surface has formed in the underlain layer of sand.

Fig. 12c: In a layer of sand or clay which is sufficiently permeable, the neutral stress lowers the geostatic stress below the groundwater surface (above this surface the geostatic stress can be regarded as increased by the capillary tension). Since groundwater lowering causes the effective geostatic stress to increase, groundwater pumping leads to the settlement of subsoils.

Fig. 13a: In the case when the subsoil is subjected to an additional load, for example, by the fill *AB*, a residual hydrostatic excess pressure arises in the layer of clay; proportionally with increasing degree of primary (filtration) consolidation of this layer (accompanied by settlement), the excess pressure decreases until the state shown in Fig. 12c is attained. A residual hydrostatic excess pressure is apt to arise whenever the state of stress in the subsoil experiences a change caused, for example, by the effect of tectonic movements, provided that the rate of the change is not so small as to allow the neutral stress to balance the level of the neutral stress in the neighbourhood. An opposite case, i.e. a residual hydrostatic negative pressure (whose maximum value is limited by cavitation) can arise when the subsoil is unloaded, for example, by excavation.

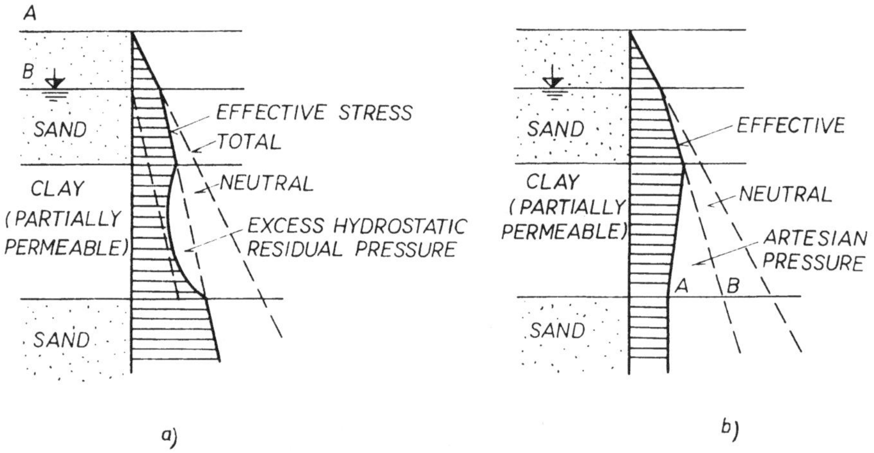

Fig. 13. Schematic diagram of vertical effective stresses in foundation soil.

Fig. 13b: If an artesian pressure *AB* acts on the bottom of a clay layer, the effective geostatic stress is further reduced by the effect of the flow pressure directed upwards, provided that the clay layer is not completely impervious (cf. Lambe and Whitman, 1969, p. 387). A reduction of the effective geostatic stress produces subsoil heaving. If the modulus of deformation of the subsoil is a function of the effective geostatic stress (a continuously non-homogeneous halfspace – cf. Section 2.5.1), rising of the groundwater surface also increases the subsoil compressibility.

In the case when neutral stresses exist, the coefficient of lateral pressure K_0 must be defined in terms of effective stresses. By eqn. (15)

$$K_o' = \frac{\sigma_x'}{\sigma_z'} \tag{37}$$

If K_0 is given in terms of total stresses, its magnitude, and hence also the magnitude of the lateral (horizontal) total stress, depends on the neutral stress as follows (Bishop, 1958; Kézdi, 1962):

$$\sigma_x = K_0'\sigma_z + (1 - K_0')\, u \tag{38}$$

In contrast to the geostatic stress, the neutral stress is isotropic (hydrostatic), defined by a spherical tensor formally identical with the scalar. Hence its measurement, in the same way as for all scalar quantities, is substantially simpler.

For locations below the groundwater surface (for example, for the sand in Fig. 12a), eqn. (36) can receive another form. If

$$\gamma_d = \frac{\gamma_s}{1 + e} \tag{39}$$

where e is the void ratio, γ_s – the specific gravity of the soil grains, γ_d – the unit weight of the dry soil, and

$$\gamma_n = \gamma_d + \frac{e}{1 + e}\gamma_w \tag{40}$$

where γ_n – the unit weight of the saturated soil, then

$$\gamma' = \gamma_n - \gamma_w = \frac{\gamma_s - e\gamma_w - \gamma_w(1 + e)}{1 + e} = \frac{\gamma_s - \gamma_w}{1 + e}$$

$$\gamma' = \frac{\delta_s - 1}{1 + e}\gamma_w$$

where $\delta_s = \gamma_s/\gamma_w$ is the relative density of the soil grains. Hence eqn. (36) can also be written in the form

$$\sigma_z' = \frac{\delta_s - 1}{1 + e}\gamma_w z \tag{41}$$

Eqn. (41) has been found suitable for determining the effective geostatic stress from measured void ratios.

1.5 The coefficient of earth pressure at rest, K_0, and its role

The importance of the magnitude of the coefficient of lateral pressure in the subsoil for problems of settlement was already pointed out by Hruban (1948). It will be shown later that it is not easy to demonstrate, for example, by means of the linear constitutive relations (13), the role which K_0, or more generally, the in situ state of stress plays relative to the vertical deformation of the subsoil (i.e. to settlement).

Assume that there exists in the subsoil a geostatic state of stress (σ_z, $\sigma_x = \sigma_y$) which is changed to $\sigma_z + \Delta\sigma_z$, $\sigma_x + \Delta\sigma_x = \sigma_y + \Delta\sigma_y$ by the application of an extensive uniform load (or generally, of a load which follows the K_0-stress path; the term stress path is defined in Section 3.3). Then, according to the third of eqns. (13) the vertical strain is given by

$$\varepsilon_z = \frac{1}{E}(\Delta\sigma_z - 2\mu\,\Delta\sigma_x)$$

and if $\Delta\sigma_x = K_0\,\Delta\sigma_z$, then

$$\varepsilon_z = \frac{\Delta\sigma_z}{E}(1 - 2K_0\mu)$$

or, on substituting for K_0 from eqn. (14),

$$\varepsilon_z = \frac{\Delta\sigma_z}{E}\left(1 - \frac{2\mu^2}{1 - \mu}\right) \tag{42}$$

In consequence of the validity of the principle of superposition, the effect of the initial state of stress in the subsoil manifests itself in eqn. (42) only in the deformation parameters E, μ. Since the stresses increase along the K_0-stress path, the vertical strain is in direct proportion to the stress increment (cf. Part 3.2).

The magnitude of the horizontal stress plays a part whenever the settlement calculations are based on the non-linear theory, the stress path of the sample (Lambe, 1967), or when introducing the effect of the neutral stress according to Skempton and Bjerrum (1957) or measuring the deformation parameters of the subsoil (since their dependence on stress is non-linear, they cannot be superposed). For the last case, one can write the equation

$$E \cong a_0\left(\sigma_z \frac{1 + 2K_0}{3}\right)^n \tag{43}$$

where E – Young's modulus (of the unsaturated subsoil), a_0 = a constant, $n \doteq \frac{1}{2}$. The equation applies to $\frac{1}{2} < K_0 < 2$ (Lambe and Whitman, 1969, p. 159). In consolidated-undrained tests, Young's modulus displays a similar dependence on the mean effective normal stress σ_{oct} (Ladd, 1964; Simons and Som, 1969). A power dependence of the initial Young's modulus on hydrostatic pressure (cf. Section 3.8) is also used when introducing a hyperbolic stress-strain relation.

For σ_z = const., the dependence of E on the lateral stress σ_x is given by eqn. (43) as a parabola. Poulos and Davis (1972) measured in laboratory the magnitude of the pseudo-elastic deformation parameters E' and μ' (Young's modulus and Poisson's ratio in terms of effective stresses – in drained tests) as functions of the magnitude of the horizontal stress. For E' they confirm in principle the validity of eqn. (43) ($\frac{1}{2} < n < 1$) and find that μ' markedly decreases with increasing horizontal stress. They come to the conclusion that the value of the in situ horizontal stress should be used in the measurements of deformation parameters.

On the basis of tests with undisturbed samples of London clay, Simons and Som (1969) suggest that the modulus of volume compressibility m_v is mainly affected by the vertical geostatic stress rather than by the magnitude of the lateral stress.

A large coefficient K_0 (> 1) causes Young's modulus to be higher in the horizontal than in the vertical direction. The application of the theory of cross anisotropy (cf. Section 2.4) usually leads to a lower stress concentration under the load, and hence also to a smaller settlement. In agreement with this, Gol'dstein (1971) draws attention to the considerable significance of the initial state of stress in overconsolidated soils, which substantially reduces the magnitude of possible settlements*). An analysis of a specific mathematical (lumped-parameter) model has shown the in situ stress to have a great effect on the load-settlement curve, even though the bearing capacity of the foundation is not influenced (Hoëg et al., 1968). The effect of K_0 on the settlement of a foundation strip was demonstrated by D'Appolonia and Lambe (1970) using the finite element method – cf. Section 3.8 and Fig. 51.

In the writer's opinion, the magnitude of the coefficient K_0 and hence also of the settlement, can considerably be affected by man's activities, for example, mining. This can explain certain anomalous settlements in undermined regions.

1.6 Unloading of subsoil produced by excavation

When a subsoil is unloaded by an extensive excavation of a constant depth d, the change in the in situ state of stress in the central portion of the excavation can be regarded as homogeneous, and the geostatic stress can be defined by the equation

$$\sigma_z = \gamma_0(z - d) \tag{44}$$

This equation is adequately approximate whenever the width of the excavation is larger than triple the width of the foundation (Sowers, 1962). Another procedure (Florin, 1959) consists of loading the surface of the subsoil by a negative load whose intensity is equal to the weight of the soil excavated from the foundation pit (Fig. 14). The in situ stress of the subsoil unloaded by an excavation is then obtained by subtracting the stress produced by the (negative) surface load q (which is usually calculated using the elasticity theory of a homogeneous and isotropic halfspace) from the in situ stress which existed in the subsoil prior to the excavation. This procedure is used, for example, by Taylor (1948, p. 281) as well as by Sowers (1962). For narrow foundation pits it is much more exact than eqn. (44).

Although the boundary conditions on the sides and the bottom of the excavation

*) The finding that during the process of construction up to 60% of the total settlement takes place in overconsolidated clays and only about 15% in normally consolidated clays (Simons, 1969) is also important.

are not fully satisfied (Fig. 14 shows the magnitude of stresses normal to the contour of a foundation pit, established by Golovin (cf. Florin, 1959), which – as for the shear stresses – should be zero), the solution is quite acceptable provided that the width of the foundation pit is equal to at least double its depth. A more complicated solution which satisfies the boundary conditions more exactly can be devised by using a certain system of fictitious vertical and horizontal forces acting on the surface of a halfspace without an excavation (Florin, 1959).

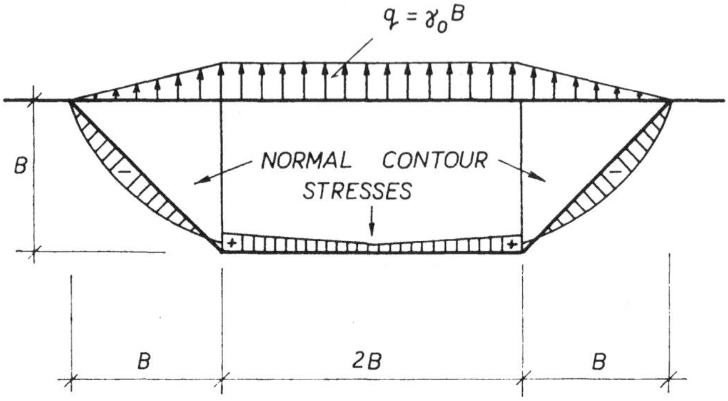

Fig. 14. Calculation of the unloading effect of excavation (Florin, 1959).

The most reliable of all is a solution by the finite element method, for by its means one can determine not only the in situ state of stress after the unloading of the subsoil by an excavation but also the displacements in the sides and the bottom of the foundation pit (Chang and Duncan, 1970). The reliability of this method is contingent to the fact that the load which replaces the weight of the excavated soil acts directly on the contour of the excavation rather than on the surface of the subsoil prior to the excavation (Desai and Abel, 1972, p. 328). Hence the boundary conditions of the solution are satisfied more precisely. Moreover, the method makes use of non-linear stress-strain relations which are more realistic. For linear elasticity, a useful solution by means of the finite difference method was presented by Dibiagio. Some of his results are quoted by Morgenstern and Eisenstein (1970). An analysis made by Christian and Wong (1973) indicates that – if the process of excavation must be approximated by a number of steps when using nonlinear constitutive equations – the number of steps used in connection with linear constitutive equations should be as small as possible (preferably just one, i.e. excavation from the surface to the final contour of the pit) since only such a procedure can guarantee a minimum of errors. As the unloading of the subsoil by the excavation of a foundation pit lasts frequently only a comparatively short time, a more exact calculation must pay heed to the time variations of the effective stress produced by gradual dissipation of the neutral negative pressure (deconsolidation of the subsoil – Feda, 1970, Šuklje, 1969, p. 371).

A practical analysis can very often be restricted to two limit cases: the undrained case (zero consolidation) and the drained case (complete consolidation). An actual solution carried out by Chang and Duncan (1970) has shown that the mean of these two extreme cases describes the in situ conditions.

1.7 Methods of measurement of in situ stresses

1.7.1 Neutral stress

The measurement of the neutral stress is the most common type of measurements performed in all subsoil explorations. The neutral stress is measured by means of either open or closed piezometers. A lined borehole represents the simplest kind of open piezometer, the neutral stress being in this case defined by the level of the groundwater surface in it. Closed piezometers designed according to different principles (a closed hydraulic circuit, electric system, etc.) measure the neutral stress directly. The same method of neutral stress measurement is used in earth structures. In contrast to in situ measurements, such cases usually involve an unsteady water motion and therefore strict requirements are imposed on the functions of piezometers, especially on their hydrostatic time lag. In in situ measurements made in undisturbed subsoils, the way in which the borehole is sealed in the vicinity of the piezometer (for example, by balls of bentonitic clays) is of as much importance as the instrument proper.

The literature on the problem of piezometric measurements is very extensive. A good review of the main problems can be found in a book by Terzaghi and Peck (1967). Cases of measurement of neutral stresses in subsoils are presented, for example, by Skempton and Henkel (1961), Gibson and Marsland (1961), Bishop et al. (1961) in the Proceedings of the Conference on Pore Pressure and Suction in Soils, which, as with the collection of papers of the International Conferences on Large Dams, bring a wealth of interesting details of piezometric measurements.

Fig. 15 reproduced from a paper by Skempton and Henkel (1961) clearly shows the complicated character of the neutral stress at various depths, resulting from the variability of the subsoil as well as from the effect of man's artificial interferences (in the case being considered, of pumping from deep wells reaching all the way to the chalk subgrade). An increase in the effective stresses produced by the effect of the flow pressure (cf. Fig. 12b for a low-permeability clay) is quite noticeable. An increase of the effective stress brought about by falling groundwater surface induced by pumping from wells (Fig. 15) is a well-known cause of the subsidence of entire cities. The most familiar case is probably the sinking of Mexico City (Carrillo, 1948) where in places the subsoil settled by more than 8 m in the course of this century (Marsal, 1975).

Certain measurements of neutral stresses made at great depths yield abnormally high values (Hanshaw and Bredehoeft, 1968). These values can be attributed not only to a rapid accumulation of sediments and to other influences but also (cf. Price, 1969) to lateral deformations of the sediments, that is to say, to an effect similar to that shown in Fig. 13a.

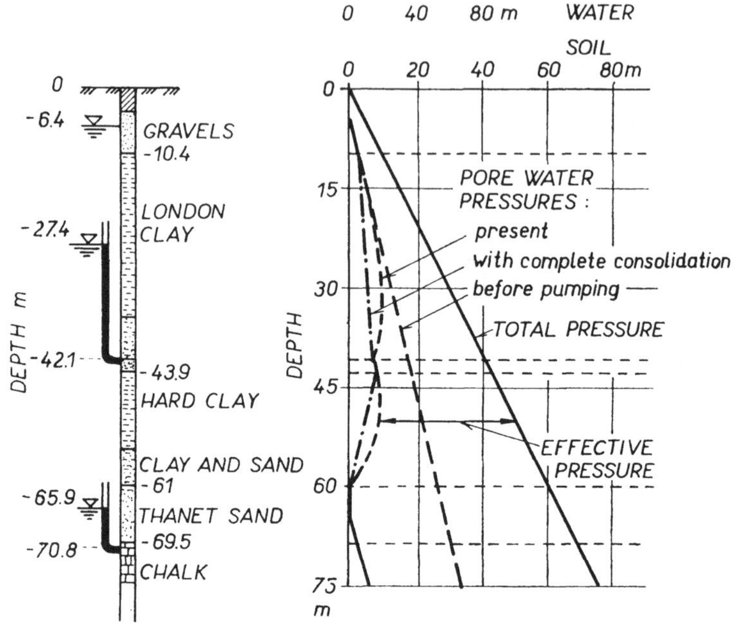

Fig. 15. Effective vertical geostatic pressure of London clay on the South Bank site (Skempton and Henkel, 1961).

1.7.2 Vertical stress

The in situ vertical stress σ_z is usually obtained by means of eqns. (35) or (41). Before applying these equations, one must determine the unit weight or the void ratio of an undisturbed sample, and the neutral stress.

A correlation between the void ratio and the effective stress can also be established by laboratory measurement of consolidated disturbed samples prepared with the original consistency on the liquid limit. A comparison of laboratory and field measurements was published, for example, by Skempton (1953) – see Fig. 16. An agreement between the laboratory and the field data is found to exist up to depths of 1000 to 1500 m (σ_z is about 10 – 15 Mpa). At greater depths, the in situ void ratio decreases at a faster rate than the laboratory one (cf. Fig. 16 showing the case of Venezuelan clayey shales). It is assumed that the foundation soil is undergoing re-crystallization by the effect of high pressures to which it has been exposed for a long time.

things, by the so-called quasi-overconsolidation effects. The structural strength,

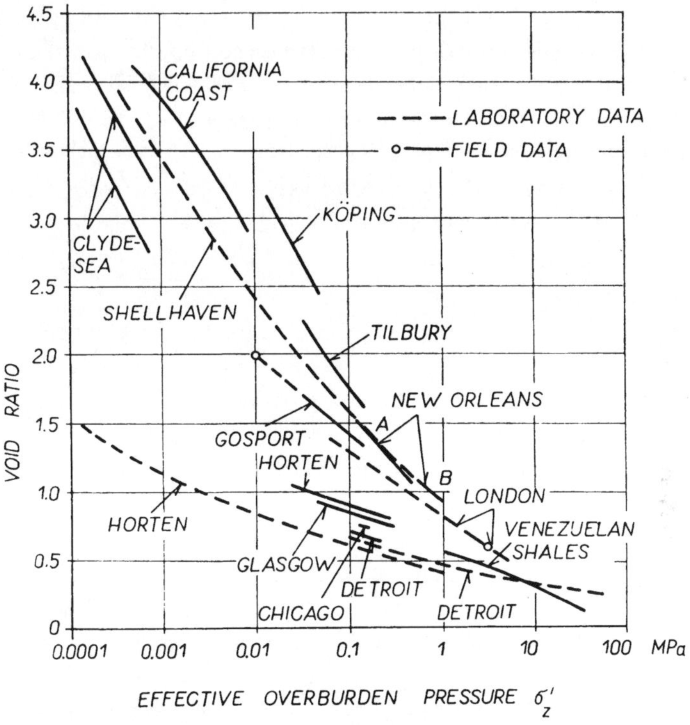

Fig. 16. Relation between the void ratio *e* and the effective geostatic pressure for various subsoils (Skempton, 1953).

whose magnitude is in fact characterized by the overconsolidation load, need not be merely a function of the previous geostatic stress σ_z' but can also depend on chemical cementation (Kenney et al., 1967), prolonged duration of so-called delayed compression (Bjerrum, 1967), desiccation, etc.

1.7.3 Magnitude of K_0'

If σ_z' is known, the magnitude of K_0' is obtained by determining the value of σ_x'; this can be done either directly or indirectly.

1.7.3.1 Direct method

In the application of this method, the value of the horizontal stress is measured directly. A method used on a large scale in rock mechanics is the unloading method, in which a part of the mass which transmits the stresses is removed at the place of measurement by boring. The ensuing decrease in stress (unloading) produces deformations in the sides of the borehole, which are measured. The value of the horizontal stress is obtained either from the measured deformations (if Young's modulus of the rock is known) or by directly measuring the pressure which is necessary to annul them. The measurements generally show the horizontal stresses to be larger by far than those corresponding to the hydrostatic state of stress (for a comprehensive information see Voight, 1966). An analogous method can be used in soils (Mencl, 1966; Zolotarev and Shariy, 1973) in which reliable measurements are much more difficult to carry out than in rocks.

A fairly important method is the direct measurement of σ_x' by means of earth pressure cells. The results obtained by this method are reported by Kenney (1967). The horizontal effective pressure was determined from the total pressure and the neutral stress measured in the walls of a hollow steel section (approximately 47 cm across). Since it was not possible to prevent completely the formation of the zone of remoulding on the section perimeter, the measured value of K_0' was smaller than the actual. In normally consolidated sensitive clays, the maximum K_0' was 0.65. The lowest values corresponded to a locality in which the surrounding subsoil experienced greatest disturbances during the process of pressing-in the gauge section. The magnitude of K_0' increased with depth (the measurements were performed at depths of 4 – 10 m). Kenney (1967) attributes this finding to the fact that the degree of disturbance of the subsoil during the penetration of the gauge section decreased with depth. If the effect of subsoil disturbance is taken into account, the in situ values of K_0' were somewhat higher than the laboratory ones (laboratory $K_0' = 0.5 - 0.6$).

Similar measurements are described by a number of authors (quoted, for example, by Wroth, 1972). The results of such measurements are frequently unfavourably influenced by the cross sensitivity of earth pressure cells of the diaphragm type (Brown, 1969, 1973).

A method recently developed for normally consolidated clays is based on hydraulic fracturing of the subsoil. The magnitude of K_0' can be determined from the magnitude of the piezometer-induced pore stress at which the crack just closes up (Bjerrum and Andersen, 1972; Bjerrum et al., 1972).

Probably the most suitable of all is the method of measuring by means of an autoforeur (self – boring) pressuremeter, developed and refined in recent years (Baguelin et al., 1975; Wroth and Hughes, 1973). This instrument disturbs the subsoil in which the geostatic stress is being measured, as little as possible. Some check measurements testify to the satisfactory function of the instrument (Jézéquel, 1973). Fig. 17 shows the results obtained by various methods. In a normally consolidated region ($z > 3$ m), eqn. (19)

is confirmed to be valid; the method of hydraulic fracturing yields larger K_0' but results of opposite nature are reported by Wroth and Hughes (1973). According to Jézéquel (1973), K_0' measured in the upper layers of the subsoil amounted to 4 and 5; for the lower-lying clays and silts $K_0' = \text{const.}$; for organic clays and peats, on the other hand, K_0' decreases with depth, and attains values as low as 0.25.

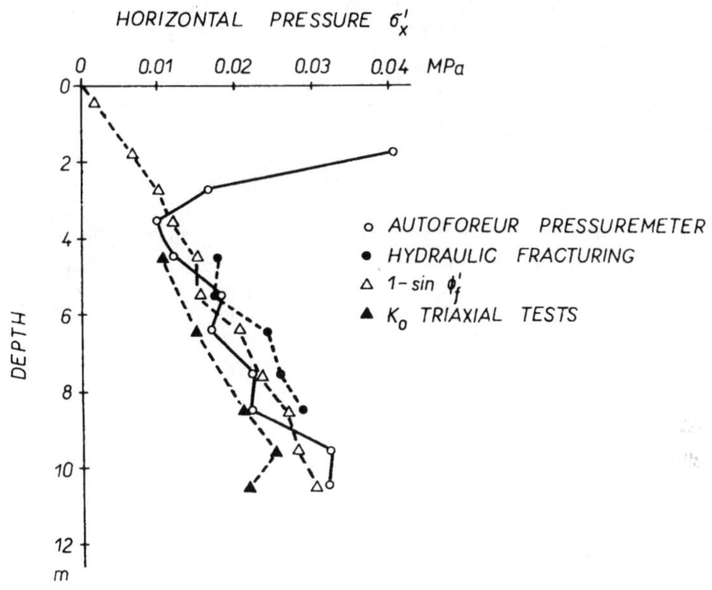

Fig. 17. Results of measurements of horizontal geostatic pressure in soft clay in Cran obtained by various methods (Baguelin et al., 1975).

1.7.3.2 Indirect method

There exist several methods of determining the magnitude of K_0' indirectly. Thus, for example, one can ascertain the *OCR* and obtain the magnitude of K_0' from relations which exist between K_0 and the overconsolidation ratio (cf. Fig. 5), according to Ladd (1965), using eqns. (23) and (24), etc.

A more exact method was described by Esu and Calabresi (1969). The magnitude of the horizontal consolidation pressure on undisturbed samples taken horizontally is determined by a procedure similar to that used in the case of vertical load, i.e. by Casagrande's method. An analogous method was developed by Poulos and Davis (1972) for normally consolidated clays: the consolidation load is determined as the cell pressure under which a change occurs in the slope of the curve defining the dependence of this pressure and the volume of the sample subjected to a constant vertical load.

The above methods are based on the assumption that overconsolidation was induced by a change in the state of stress of the subsoil, without the effect of other structural actions (for example, cementation of solid particles). If this assumption is not satisfied then – an ascertained overconsolidation load notwithstanding (the so-called quasi-overconsolidation) – the actual value of K_0' is apt to correspond to that of a normally consolidated subsoil (cf. Burland, 1971).

The method employed by Skempton (1961) in the reconstruction of the geotechnical history of London clay at Bradwell, can also be regarded as indirect. The value of K_0' is determined from the magnitude of the capillary pressure p_k and the coefficient of pore pressure A_s using the relation (Skempton, 1961)

$$p_k = \sigma_z'[K_0' - A_s(K_0' - 1)] \tag{45}$$

The value of A_s is measured in the laboratory on undisturbed samples (it is usually 1/3) as for the value of p_k which was ascertained by the following four methods: direct and indirect (with interpolation) measurement of pressure which stops swelling of the sample, direct measurement of suction of the sample, and an analysis of undrained strength of the sample. Since the results of all four methods agreed reasonably well, K_0' could be determined using the average value of p_k and eqn. (45). Its magnitude ranged from $K_0' = 1.5$ (at a depth of about 30 m) to $K_0' > 2.5$ in the upper 10 m (Fig. 7). The same method was used by Bishop et al. (1965).

Among the indirect methods one can also include the following version of the direct unloading method: an earth or a rock mass is unloaded by means of a gallery, and measurements are taken of the displacements of the contour of the gallery. The deformation parameters of the medium established by independent field tests, are used in the finite element method to calculate the contour stresses in the gallery such as will eliminate the measured displacements. These stresses give the geostatic state of stress of the medium.

This procedure was applied to the Ordovician stratified shales of Prague's bedrock (Doležalová et al., 1975). The geostatic stresses were calculated from the measured displacements of the contour of a circular gallery (of 7.5 m dia.). The calculation assumed a linear medium with deformation isotropy (the deformation anisotropy was found to be insignificant) and strength anisotropy. The coefficient K_0 was calculated in three profiles spaced 2 m apart; its gradual decrease from the drift end of the gallery (A: $K_0 = 0.32$; B: $K_0 = 0.69$; C: $K_0 = 0.94$) seems to indicate the ever growing extent of disturbances around the gallery. Hence the first profile, A, is in a state which is closest to the actual geostatic state of stress. The calculated displacements ranged between 50% and 100% of the measured ones. The circumstance that the deformation parameters were measured during a short-term (one-day) test, played an unfavourable role; short-term deformations of the gallery took two weeks and were evidently visco-elastic.

Chapter 2

SUBSOIL STRESSES PRODUCED BY EXTERNAL LOAD

2.1 Introduction

The first step in the determination of the state of stress in a foundation soil, is the ascertainment of the character of the external load. The term "the character of the external load" is understood to mean:

(a) the intensity and the geometry of the external load, i.e. whether the load is concentrated or continuous, and in the latter case, what is the law of its distribution (for example, uniform, triangular, trapezoidal, etc.), its direction and inclination and whether it is applied to the surface or at some depth;

(b) the dynamics of the external load, i.e. the way in which the intensity and the geometry of this load vary with time.

An exact determination of the load character by means of stochastical relationships is frequently either impossible or extremely tedious. That is why, in engineering calculations, the load character is defined by means of so-called load combinations which can be classified, for example, as follows:

– basic combinations consisting of a constant, long-term variable and two or more short-term loads or effects,

– extraordinary combinations consisting of a constant, long-term, random and some short-term and extraordinary loads and effects.

It is assumed that in a halfspace (foundation soil) the total stress propagates at a virtually infinite rate (compared with the rate of variation of the usual types of loads for which the settlement is calculated*). This does not, however, apply to the effective stress whose variability with time is caused by the neutral stress. The neutral stress is time dependent because the volume changes of a saturated soil produced by the load, give rise to an unsteady filtration field in its pores. Under loads acting during short time intervals (wind, crane loads, etc.) the resulting volume deformations are

*) It is physically justified to regard this propagation as a diffusion process (cf. Section 2.2.2). In an actual foundation soil the total stress propagates at a rate of several hundreds to thousands of metres per second. At a load moving with the velocity of sound, shock effects are apt to begin to manifest themselves in the subsoil. In this study, the rate of loading is considered to be so low as to make the dynamic effects negligible.

usually negligible (the effective stress can be assumed to be constant), and their effect is only manifested by a change in the shape of the foundation soil elements. Hence only the constant and long-term variable load is considered in the calculations of the consolidation settlement (i.e. settlement produced by volume changes of the subsoil).

In the calculations of the state of stress in the subsoil, the real soil is replaced by a mathematical model of a halfspace which can be endowed with different mechanical properties. The fewer and simpler are the mechanical parameters (for example, two parameters of the isotropic theory of linear elasticity), the more straightforward is the calculation but the greater can be the deviation of the behaviour of the mathematical model from that of the real medium. With increasing number of parameters (for example, twenty-one parameters of a generally anisotropic elastic halfspace), on the other hand, the calculations and the test procedures used for measuring these parameters become so complicated as to be unfeasible. The calculations of the state of stress, therefore, start reasonably from the simplest model, while somewhat more complicated models (for example, a cross-anisotropic halfspace) are only selected in cases when substantial differences between theory and reality might be expected.

The methods of calculating the state of stress produced by an external load (a halfspace, except a two-phase halfspace, is usually assumed to be weightless) can be classified in various ways. The most essential point in which the methods differ is the supposed continuity or discreteness of the medium. The assumption about discreteness is so complicated that neither possible non-homogeneity, nor anisotropy, etc. are additionally introduced, although this can theoretically be done.

Solutions based on the assumption of a continuum can be grouped according to further hypothetical qualifications of the halfspace in question (Fig. 18). Stress calculations can also be classified from a mathematical point of view, i.e. whether they lead to a closed form (e.g. elastic halfspace) or an open form (numerical methods, for example, the finite element method) solution, etc.

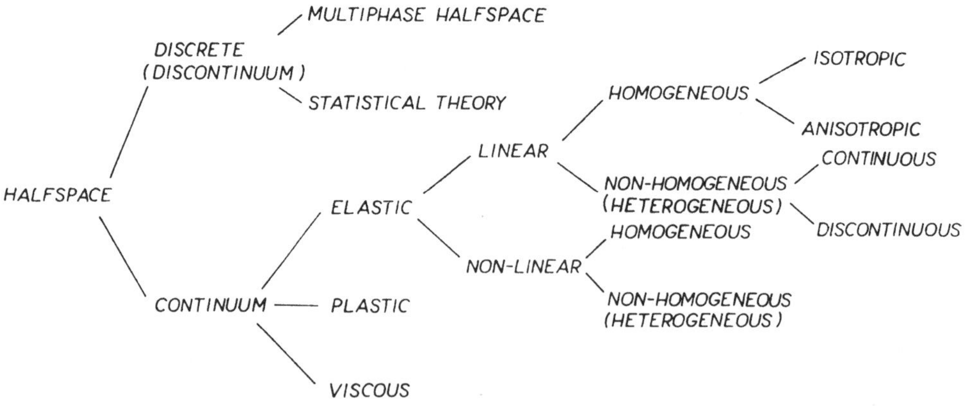

Fig. 18. Classification of the methods of stress calculation.

The writer adopts the classification shown in Fig. 18. His considerations do not include the states of stress in plastic, elastoplastic*) and other halfspaces, and in halfspaces with viscous properties (visco-elastic, visco-plastic, etc.) which, in explicit conception, lie outside the scope of this book. Implicitly, however, the state of stress in the subsoil is always of an elasto-plastic nature. For the sake of brevity, the adjective "elastic" is left out of the titles of the subsequent sections.

A comprehensive review of practically all published solutions relating to stresses in an elastic continuum, including a solution of an orthorhombic and cross-anisotropic medium due to Gerrard and Harrison (1970a,b) and some of Giroud's solutions (1972, 1973), is presented by Poulos and Davis (1974).

2.2 Discrete media

In stress calculations the idea of the discrete nature of a medium is respected in one of the following ways: the halfspace is regarded as multi-phase, and the stresses are determined for each phase separately (usually only a two-phase halfspace is considered, i.e. foundation soil saturated with water). Alternatively, the halfspace is taken to be a particulate medium in which stress propagation obeys statistical laws.

The third method of calculating stresses in a discontinuous medium (a rock mass) is based on the finite element technique. The discrete nature of the medium is taken into account by making the resulting stiffness matrix composed of the stiffnesses of the continuous elements and of the stiffnesses of the joint elements (Goodman et al., 1968). The method was already mentioned in Section 1.2.3. The particulate character of this medium manifests itself – among others – by the medium's dilatancy under certain conditions (Goodman and Dubois, 1972). An assumption explicitly introduced by the method is that the joint elements do not possess any normal tensile strength. Because of its specific character, the method is discussed at some length in Section 3.8.

2.2.1 Two-phase halfspace

As a sound approximation, one regards the assumption that the total stress in a saturated subsoil (clay, as a rule) produced by an external load propagates according to the laws applicable to a continuum, viz. an elastic, linear, homogeneous and isotropic halfspace. This assumption implies the belief that the inaccuracy of the subsequent calculation procedures is so great that the use of a more complicated model of the halfspace plays only a minor part in the final result. One either assumes that the halfspace is incompressible (Poisson's ratio $\mu = 0.5$) and obtains the magnitude

---

*) A limit state of stress exists at some of their points or regions.

of the neutral stress at the instant of a load increase, or – using the theory of filtration consolidation – determines the variation of the neutral stress with time. In view of the aim of this monograph, the writer will deal only with the former case.

Essentially, two procedures are available for the purpose. In one, the calculated increments (or decrements, in the case of unloading) of the total stress are applied in laboratory to an undisturbed sample of the foundation soil (usually in a triaxial apparatus) in a way to make the stress path (the line connecting successive states of stress each of which is graphically represented by a stress point in the stress space) in the laboratory sample coincide with that of the in situ point being considered. The increment of the neutral stress is measured directly. The applications and refinements of the method are most comprehensively discussed by Lambe (Lambe, 1964, 1967; Lambe and Whitman, 1969, p. 403 (the example)). See also Section 3.3.

In the other procedure the increment of neutral stress Δu is calculated by means of the Skempton equation (Skempton, 1954)

$$\Delta u = B[\Delta\sigma_3 + A(\Delta\sigma_1 - \Delta\sigma_3)] \tag{46}$$

where the principal stresses $\sigma_1 = \sigma_z$, and $\sigma_3 = \sigma_x = \sigma_y$. For two-phase systems, $B = 1$ and A is determined on an undisturbed sample in the laboratory (for $A = B = 1$, $\Delta u = \Delta\sigma_1$, which is the assumption of the uniaxial theory of consolidation).

The second procedure is in fact a less perfect substitute of the first method in the case when the in situ stress path is hard to simulate exactly in the laboratory. This fact notwithstanding, the laboratory stress path must resemble the in situ one as closely as possible during the measurement of the parameter A. Examples of this procedure are presented, for example, by Gibson and Marsland (1961), Lambe and Whitman (1969, p. 401) and by Hoëg et al. (1969). The last named authors used eqn. (46) in its generalized form for $B = 1$ (cf. Skempton, 1960; Henkel, 1960)

$$\Delta u = \Delta\sigma_{oct} + 3\alpha\,\Delta\tau_{oct} \tag{47}$$

where

$$\Delta\sigma_{oct} = \tfrac{1}{3}(\Delta\sigma_1 + \Delta\sigma_2 + \Delta\sigma_3)$$

$$\Delta\tau_{oct} = \tfrac{1}{3}[(\Delta\sigma_1 - \Delta\sigma_2)^2 + (\Delta\sigma_2 - \Delta\sigma_3)^2 + (\Delta\sigma_3 - \Delta\sigma_1)^2]^{1/2}$$

and assuming that $\alpha = 0$ (i.e. using the assumption of the elasticity theory – volume changes are not influenced by the stress deviator) they obtained

$$\Delta u = \Delta\sigma_{oct} \tag{48}$$

The study of Hoëg et al. (1969) is remarkable for the fact that it suggests the possibility of calculating Δu also for an elasto-plastic halfspace, i.e. for a foundation soil that has locally reached its shear strength. The authors start from an assumption (verified in a study of Hoëg et al., 1968) that $\Delta\sigma_1$ can be determined by means of an analysis according to the elasticity theory even after local plastic deformations have been initiated. Since, however, $\Delta\sigma_1 - \Delta\sigma_3 = \text{const.}$ (the shear strength inside a local plastic region does not increase), it must hold in a plastic region that $\Delta\sigma_1 = \Delta\sigma_3$ for stress increments which have appeared since the onset of the plastic state. Using eqn. (48) one can then extrapolate the calculation for this state.

This physically justified procedure shows a striking difference from the procedure of Lambe (1970) who based the prediction of pore pressures in the subsoil of a fill on the elasticity theory although the calculated shear stresses were twice to three times as large as the strength of the subsoil. Despite a comparatively good agreement between the measured and the calculated pore pressures, a physically ungrounded mathematical model cannot be considered to be reliable for prediction purposes – "correct prediction does not prove that the technique is correct" (Lambe, 1970).

The calculations of Hoëg et al. (1969) relating to the elastic as well as to the plastic range, and those of the other above-named authors are in excellent agreement with in situ measurements (see also D'Appolonia et al., 1971). The increase in pore pressures resulting from the plastification of the subsoil (collapse of clay structure) was also described by other researchers (e.g. Chang et al., 1973).

Reliable predictions of neutral stresses can also be made using the Cambridge theory of the state boundary surface (see Burland, 1971; Wroth, 1977).

2.2.2 Statistical theory

The logic of the statistical approach to stress analysis of foundation soils was first demonstrated by Kögler and Scheidig (1928). They start out from the idea that the foundation soil consists of a set of mutually contacting grains (spheres). The pressure acting on a grain is transmitted to adjacent grains with which the grain is in contact, and this mechanism leads to a bell-shaped pressure distribution which closely resembles an experimental distribution (Kögler and Scheidig, 1938, pp. 77 and 79). This theory was elaborated by Pokrovskiy (1937), and after him, mainly by Kandaurov (1959, 1960, 1966) and Smoltczyk (1967).

The problem can be formulated more generally. The propagation of stresses in a halfspace can be regarded as a diffusion process described by stochastical relationships. The basis of these relationships is formed by the Chapman-Kolmogorov equation which was first derived by Smoluchowski in connection with the theory of Brownian movement. Its solution leads to a partial differential equation of the parabolic type, familiar in soil mechanics in the theory of filtration consolidation. The stochastical approach was used by Litwiniszyn (1957, 1964) in the solution of a problem of deformation of a granular medium, and by Muller (1962) in the solution of the state of stress. In both cases the procedure leads to a normal Gauss-Laplace distribution function.

Kandaurov (1966) replaces the real subsoil by a medium consisting of grains of finite dimensions (granular medium), and assumes that in a plane problem, the stress of one particle is transmitted to two other particles, and in a three-dimensional problem, to four particles. The distribution of stresses produced by a concentrated force in each plane z = const. follows the binomial law which can be approximated by the normal frequency curve.

The resulting formulae of Kandaurov (1966) and Smoltczyk (1967) have the following essential features in common:

(a) The components of the stress tensor have a character of random quantities with a normal (Gauss-Laplace) distribution*).

(b) The magnitude of the stresses (including the stress σ_z) depends on a special parameter which, according to Kandaurov (1960, 1966), characterizes "the distribution ability of the medium" (it can alternatively be expressed in terms of the so-called coefficient of non-uniformity of transmission, or more suitably, in terms of the coefficient of lateral pressure η which can be measured in laboratory – Kandaurov, 1966, pp. 138, 140, 146); Smoltczyk calls this parameter "the concentration parameter".

This parameter cannot be determined in advance, and this fact can be qualified as a drawback of these theories (see also Tsytovich, 1963, p. 307). If – in keeping with Kandaurov (1966, p. 146) – $\eta = \sigma_x/\sigma_z$, η cannot be taken to be a mechanical parameter of the foundation soil. The condition η = const. is only a special case of the state of stress in the subsoil, for example, for $\varepsilon_x = 0$ (then $\eta = K_0$) when the stress path at each point of the subsoil is radial**) for the stresses σ_x and σ_z.

A comparison of the distributions of the vertical normal stress σ_z calculated by means of the elasticity theory (Boussinesq) and by means of the statistical theory according to Smoltczyk (qualitatively similar results are also obtained by means of Kandaurov's theory) for a halfspace subjected to a vertical concentrated load P, is shown in Fig. 19. For the purpose of this comparison, the magnitude of the concentration parameter was chosen so as to give the same maximum values of σ_z for both solutions (according to eqn. (49), this step demands that $\eta = 0.33$; cf. Kandaurov, 1966, p. 289).

The difference between the two distributions of σ_z is small. On the strength of this fact Smoltczyk (1972) believes that the magnitude of σ_z is nearly statically determinate, i.e. that it can be obtained from the conditions of equilibrium, and that its dependence on a particular form of the constitutive equations is only secondary. As will

---

*) Thus, for example, Kandaurov (1966, pp. 140, 167, 287) defines the stress σ_z from a vertical concentrated load P, by the relation

$$\sigma_z = \frac{P}{2\eta\pi z^2} \exp\left(-\frac{r^2}{2\eta z^2}\right); \tag{49}$$

for notation, see Fig. 20; η — the coefficient of lateral pressure.

**) A radial stress path in the space or in the plane of stress is a straight line passing through the origin of the space or the plane.

be shown later, this applies only to a homogeneous, isotropic and linear halfspace, loaded on its surface.

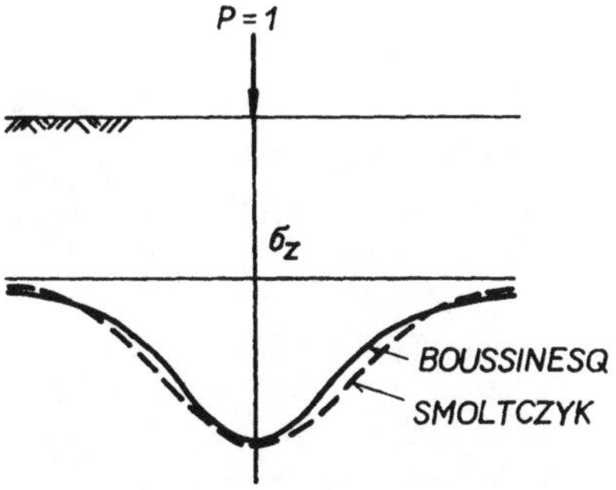

Fig. 19. Values of stress σ_z obtained by means of the elastic (solid line) and the statistical (dashed line) theory (Smoltczyk, 1967).

Although the writer does not regard the statistical theory as exceedingly important quantitatively*) (some of its interesting applications are described by Smoltczyk, 1967), he attributes to it a considerable qualitative importance. The statistical theory provides a noteworthy picture of the physical mechanism which controls the propagation of stresses in foundation soils. The use of the diffusion theory represents an important synthetic step which unifies the stress theory with the theory of filtration consolidation (and other mathematically analogous processes).

2.3 Homogeneous isotropic and linear continuum

The solutions of the state of stress in this model of halfspace constitute, both historically and methodologically, the basis of all solutions of stresses in a continuum. Depending on whether the load is applied to the surface, inside the halfspace or inside the space, the solutions are divided into three groups. The first case is indisputably the most important.

The solution of the state of stress in this continuum goes into considerable detail, and it is discussed at length by Florin (1959), Scott (1963), Tsytovich (1963), Kézdi (1964), Myslivec et al. (1970), Giroud (1972, 1973), Poulos and Davis (1974), and others. The studies of these authors contain far more particulars than will be presented in this work.

2.3.1 Surface loading of a halfspace

A fundamental solution of this problem, i.e. of stresses in a horizontally bounded halfspace loaded by a concentrated force, was presented by Boussinesq (1885) and

*) From the design standpoint, it can at most be considered to be equivalent to the continuum theory. Kandaurov et al. (1969) use the two theories side by side.

Cerruti (1882), the former author obtaining it for a vertical load, the latter for a load of general direction*).

Boussinesq (1885) started from the conditions of equilibrium formulated in terms of the displacement components. He derived the state of stress from the strain of the halfspace thus obtained. The displacements are expressed as functions of the potential: ordinary (for boundary conditions given by the displacements); logarithmic of the first kind (for a vertical surface loading); and logarithmic of the second kind (for a shear, i.e. horizontal surface loading).

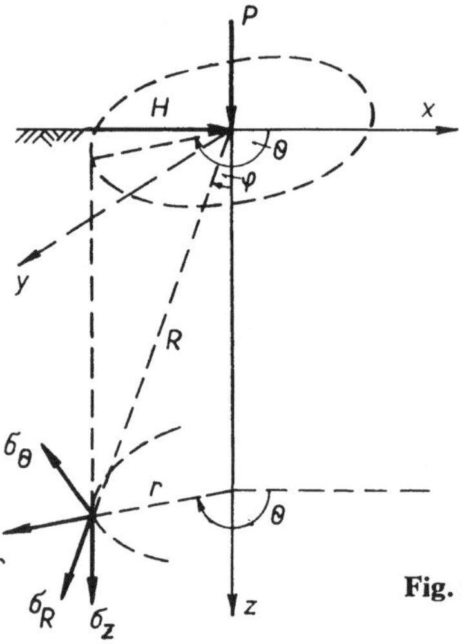

Fig. 20. Cylindrical and Cartesian (rectangular) coordinates.

Using the cylindrical coordinates (which are convenient because the problem is axisymmetrical) shown in Fig. 20, the result of his solution, i.e. the stress components, can be written as follows:

$$\sigma_z = \frac{3P}{2\pi}\frac{z^3}{R^5}$$

$$\sigma_r = \frac{P}{2\pi z^2}\left[3\cos^3\varphi\sin^2\varphi - (1-2\mu)\frac{\cos^2\varphi}{1+\cos\varphi}\right]$$

$$\left(\text{or}\quad \sigma_r = \frac{P}{2\pi R^2}\left[\frac{3r^2 z}{R^3} - (1-2\mu)\frac{R}{R+z}\right]\right)$$

$$\sigma_\theta = -(1-2\mu)\frac{P}{2\pi z^2}\left[\cos^3\varphi - \frac{\cos^2\varphi}{1+\cos\varphi}\right]$$

*) Before publishing his book, Boussinesq reported the main results of his solution for a vertical load in C.R. Acad. Sci. in 1878—9 (vol. 86, 87 and 88). In 1885, stimulated by Cerruti's work, he supplemented his results to include the effect of a load acting in general direction; his procedure is, however, essentially different from Cerruti's.

$$\left(\text{or} \quad \sigma_\theta = -(1-2\mu)\frac{P}{2\pi R^2}\left(\frac{z}{R} - \frac{R}{R+z}\right)\right)$$

$$\tau_{rz} = \frac{3P}{2\pi z^2}\cos^4\varphi \sin\varphi$$

$$\left(\text{or} \quad \tau_{rz} = \frac{3P}{2\pi}\frac{rz^2}{R^5}\right) \tag{50}$$

(for reasons of symmetry, $\tau_{r\theta} = \tau_{z\theta} = 0$, σ_θ is the principal stress); the + sign is affixed to compression, the − sign to tension, μ – Poisson's ratio. The derivation of these relations is also given by Love (1927), Timoshenko (1934), Lur'ie (1955) and others.

Cerruti's (1882) procedure consisted of direct integration by means of Betti's "reciprocal theorem" method (cf. Love, 1927). Cerruti is usually quoted as the author of the solution of the state of stress in a halfspace subjected to a horizontal (shear) force H (Fig. 20)*) because this part of his solution is antecedent to Boussinesq's solution. Since no axi-symmetrical load is involved, Cerruti's solution is expressed in rectangular (Cartesian) coordinates (x, y, z) with a spherical R in the form

$$\sigma_x = \frac{H}{2\pi}\frac{x}{R^3}\left[\frac{3x^2}{R^2} - \frac{(1-2\mu)R^2}{(R+z)^2}\left(1 - \frac{y^2(3R+z)}{R^2(R+z)}\right)\right]$$

$$\sigma_y = \frac{H}{2\pi}\frac{x}{R^3}\left[\frac{3y^2}{R^2} - \frac{(1-2\mu)R^2}{(R+z)^2}\left(3 - \frac{x^2(3R+z)}{R^2(R+z)}\right)\right]$$

$$\sigma_z = \frac{3H}{2\pi}\frac{xz^2}{R^5}$$

$$\tau_{xy} = \frac{H}{2\pi}\frac{y}{R^3}\left[\frac{3x^2}{R^2} + \frac{(1-2\mu)R^2}{(R+z)^2}\left(1 - \frac{x^2(3R+z)}{R^2(R+z)}\right)\right]$$

$$\tau_{yz} = \frac{3H}{2\pi}\frac{xyz}{R^5}$$

$$\tau_{zx} = \frac{3H}{2\pi}\frac{x^2z}{R^5} \tag{51}$$

By integration and superposition (since the medium is linear) one can determine from eqns. (50) and (51) the state of stress in a halfspace for any surface loading. Since this integration is difficult, a number of fairly common cases have been worked

*) Florin (1959, p. 137) and after him Tsytovich (1963, p. 218) attribute this solution to other, later authors.

out in table form or graphically. An electronic computer can also be used to advantage for this purpose (thus, for example, Tortorici and Umiltà, 1969, propose a procedure for calculating stresses beneath several parallel foundation strips acting at the same time; a study of Kandaurov et al., 1969, is a detailed design manual for project engineers, describing the calculation of the state of stress, settlement, critical and limit load, etc. in ALGOL, usually for an elastic halfspace as well as for a granular medium), for the tables of stresses are usually prepared with its aid.

The most frequent is the case of a rectangular area subjected to a load of linearly increasing intensity. This case was solved wholly generally and in a comprehensive way by Giroud (1969, 1970a, 1973); the results of his studies are set up in tables and therefore convenient for practical use.

A scheme of a linear contact load distributed over a rectangular area is shown in Fig. 21. Giroud determines the magnitude of the three normal stress components, σ_x, σ_y and σ_z beneath the centre and the corners of the loaded rectangle. A general linearly

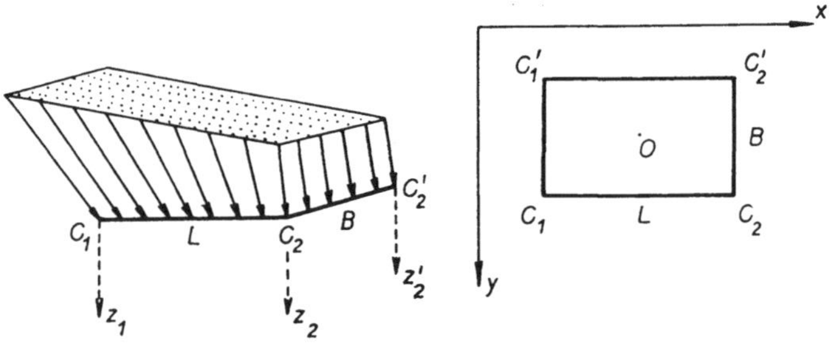

Fig. 21. Scheme of a linear distribution of contact load on a rectangle (Giroud, 1970a and 1973).

distributed load can be decomposed into four elementary types of load: q – uniform normal load, p – antisymmetric normal load, s – uniform shear load, and t – antisymmetric shear load (Fig. 22). The solution is confined to the case when the shear load is parallel to the side L of the rectangle (Fig. 21).

The resulting relations are as follows:

for the corners of the rectangle

$$\sigma_x = q[K_2 - (1 - 2\mu) K'_2] + \varepsilon p[M_2 - (1 - 2\mu) M'_2]$$
$$- \varepsilon s[K_3 - (1 - 2\mu) K'_3 - t[M_3 - (1 - 2\mu) M'_3]$$
$$\sigma_y = q[L_2 - (1 - 2\mu) L'_2] + \varepsilon p[N_2 - (1 - 2\mu) N'_2]$$
$$- \varepsilon s[K_5 - (1 - 2\mu) K'_5] - t[M_5 - (1 - 2\mu) M'_5]$$
$$\sigma_z = qK_0 + \varepsilon pM_0 - \varepsilon sK_1 - tM_1 \; ; \qquad (52)$$

for the centre of the rectangle

$$\sigma_x = 4q[K_2 - (1 - 2\mu) K_2'] + 2t[K_3 - M_3 - (1 - 2\mu)(K_3' - M_3')]$$

$$\sigma_y = 4q[L_2 - (1 - 2\mu) L_2'] + 2t[K_5 - M_5 - (1 - 2\mu)(K_5' - M_5')]$$

$$\sigma_z = 4qK_0 + 2t(K_1 - M_1) \tag{53}$$

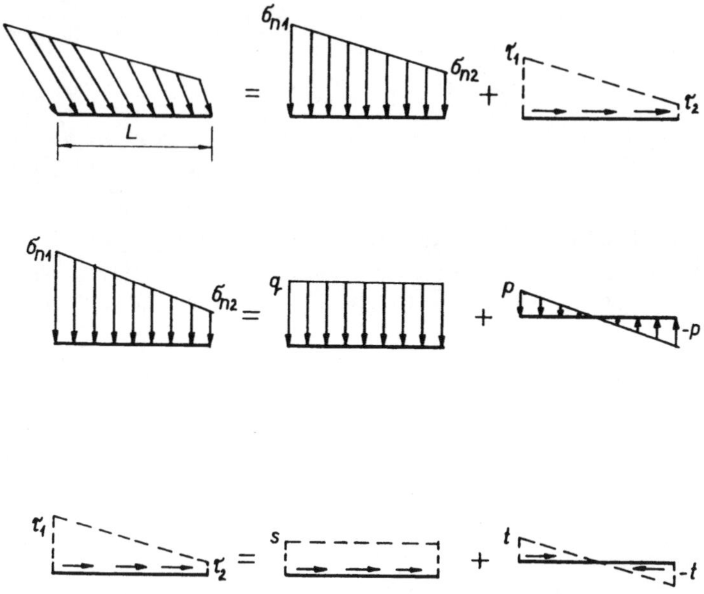

Fig. 22. Resolution of linearly distributed contact pressure (Giroud, 1970a and 1973).

In keeping with eqns. (52) and (53), the normal stresses σ_x, σ_y and σ_z depend on: Poisson's ratio $0 \leqq \mu \leqq 0.5$, on the coefficient ε ($\varepsilon = +1$ in the corners C_1 and C_1' to which correspond σ_{n1} and τ_1, and $\varepsilon = -1$ in the corners C_2 and C_2' to which correspond σ_{n2} and τ_2), and on a total of twenty coefficients K, L, M, N which are evaluated in tables and whose magnitude depends on the ratio between the sides of the rectangle, B/L (Fig. 21) and the ratio between the depth and the length, z/L (z/L for the corners, $2z/L$ for the centre, and z/L in both cases for the load p). Giroud's tables are reproduced in the supplement to this study (Tables I to XX); they are also given by Poulos and Davis (1974).

The procedure outlined above cannot be applied when the load is not distributed linearly and the loaded area is not rectangular (formulae for circular loaded areas are proposed, for example, by Scott, 1963; Florin, 1959; Kézdi, 1964; Myslivec et al., 1970; Giroud, 1972, and others). In cases of that sort one can proceed – using Saint-Venant's principle – in one of the following three ways:

(a) The actual loaded area is replaced by a rectangle of equal area, and the actual load by a linear load with equal resultant. Except for a small, locally confined region beneath the base of foundation, the stresses thus calculated will be close enough to the exact values. Thus, for example, a uniformly loaded rectangle can be replaced by a square of equal area, and the actual vertical stress will differ by less than 5% from the stress calculated at a depth equal to half the foundation width, provided that the length of the rectangular foundation is not greater than double its width (Sowers, 1962). At greater depths, the actual load can be replaced by its resultant (Taylor, 1948). This is the principle of the second procedure which will be discussed below.

(b) The actual loaded surface is divided into a number of small, usually square areas, and the load applied to them is replaced by concentrated loads. The resulting stress is the sum of stresses produced by these concentrated loads. To attain an accuracy better than 95%, at a given depth, the dimensions of the areas (squares, in the case being considered) must be smaller than 1/3 of the depth (Taylor, 1948; Florin, 1959; Myslivec et al., 1970, etc.). At a lesser depth the approximate stress along the foundation axis is higher and near the edges lower than the exact value (Giroud, 1972, p. 74).

(c) The third procedure is based on the sector method. The loaded area of any shape is divided into sectors with a common vertex (a point through which passes the vertical line along which stresses or strains are to be determined), uniformly loaded and with small vertex angles. The effect of a sector is found by integration of the concentrated loads which replace the uniform load within the range of the sector (Poulos and Davis, 1974, p. 13).

Another procedure applicable to loads uniformly distributed over irregular areas, when only the vertical stress σ_z is determined, operates with the Newmark's diagram (Newmark, 1942; see also Sowers, 1962; Terzaghi and Peck, 1967; Šuklje, 1969; Poulos and Davis, 1974, and others). The diagram consists of a system of concentric circles and radial lines which divide the loaded surface into annular sectors. The uniform load which acts on each of these sectors, contributes in the same proportion to the vertical stress at a given depth (hence a separate diagram is needed for each depth); the diagram is thus clearly one of the influence surface. The loaded area must be drawn to the scale of depth, which is marked on each diagram, and the centre of the diagram must be placed at the point through which passes the vertical line along which the stress is being determined. The resulting stress is the product of the uniform load, the number of the areas into which the loaded surface is divided, and the influence value of the diagram.

The procedures outlined above can be employed, for example, when — owing to the effect of stiffness and roughness of the foundation surface — neither the normal nor the shear component of the contact load is distributed linearly. The tests made by Wibel and Blinde (1971) show that under usual loads, the contact stress can be regarded as uniformly distributed in loose to medium dense sands while the theoretical saddle-shaped (concave) distribution should only be considered for dense sands.

The effect of the contact shear stress on the state of stress in the foundation soil (and hence usually also on the settlement of the foundation) is ordinarily regarded as negligible (Giroud, 1970b). Prange's linear analysis (Prange, 1965) clearly overestimates this effect – according to it, the contact shear stress always acts on the foundation in the direction towards its centre (centreline). It was found on the basis of a more exact analysis of an actual case (Wibel, 1971) that this only applies when $\mu < 0.38$; for $\mu > 0.38$, the contact shear stress acts in opposite direction, and for $\mu = 0.38$, it is zero. A more reliable non-linear calculation has shown a change of the first type of stress (towards the centre) into the second type (away from the centre) which proceeds in the direction from the boundary towards the centre of the foundation. Hence the effects of the contact shear stresses of different senses can compensate one another. The negligible effect of foundation roughness on the state of stress in the subsoil, particularly in the case when $\mu > 0.3$, was evaluated numerically by Yamaguchi et al. (1968) for a strip, and by Carrier and Christian (1973) for a circle.

As eqns. (50) and (51) imply, of the normal stresses only the stress σ_z is independent of Poisson's ratio, i.e. of the mechanical parameters of the foundation soil. Since in soil mechanics the measurement of Poisson's ratio is quite difficult (it depends, among others, on the stress level), this fact is particularly important from the practical standpoint.

An incompressible halfspace when $\mu = 0.5$, is a special case, for then, $\sigma_\theta = 0$ in eqns. (50) and the resultant of the stresses σ_r and σ_z passes through the point of action of the vertical load P. Hence the resulting stress is σ_R, and the state of stress is uniaxial. This is the case of the so-called radial propagation of stresses.

For a vertical line load along the y axis (Fig. 20), just as in the calculation of stresses beneath an infinitely long strip which can be deduced from the former case, the strains are two-dimensional. Combining eqns. (1) and (13) (which, in a general case when x, y, z, are not the principal axes, must be supplemented by the relations $\tau_{xy} = G\gamma_{xy}$, $\tau_{yz} = G\gamma_{yz}$, $\tau_{zx} = G\gamma_{zx}$) one can show that the solution of plane strain is identical with the solution of the plane state of stress (see, for example, Kézdi, 1964), the only difference being the magnitude of the elasticity constants (cf. Florin, 1959, p. 142). Analogous problems solved for plane strain can be converted to problems of plane stress by introducing the equivalent Young's modulus $(1 + 2\mu)\, E/(1 + \mu)^2$ and the equivalent Poisson's ratio $\mu/(1 + \mu)$; similar relations can be established for converting the problem of plane stress to that of plane strain. So long as the stress relations contain no elastic parameters, the solutions for plane stresses are identical with those for plane strains (for example, stresses in a surface loaded semi-infinite plate and stresses under a line load – Poulos and Davis, 1974, p. 9). It can readily be shown that in plane problems, no stress component depends on Poisson's ratio μ, provided the load is a surface load and no boundary condition is specified in terms of the displacements. This finding should be borne in mind when solving the state of stress in a halfspace under a line surface load.

The state of stress under a vertical line load of a halfspace along the y axis (Fig. 20) is a special case of the general equations (50); hence, it can clearly be deduced by integrating along the straight line y. Since, however, the resulting expressions must not contain μ, one has to use such forms of eqns. (50) in which μ does not appear, i.e. the formulae applicable to the case of $\mu = 0.5$. Thus for line loads the propagation of stresses in the (x, z) plane is always radial; this, however, is general in this case not contingent to a definite value of μ because it is dictated by the boundary conditions of the problem (Fröhlich, 1934, pp. 18 and 178). One can also say that the choice of $\mu = 0.5$ represents a way of introducing the condition of plane strain in the general equations (50).

Thus the stresses under a line load, and in turn, other plane problems can be obtained using eqns. (50) for $\mu = 0.5$ (Fröhlich, 1934, p. 16 et seq.). As McNamee and Gibson (1960) have shown, the solution of axi-symmetrical problems can easily be deduced from the solution of plane strain problems.

Since the solution of plane problems is attractive from the mathematical point of view, its significance is frequently overestimated. The writer excludes them from further discussion for the following two reasons:

(a) The solution is a special case of eqns. (50) and (51) and can unconditionally be derived from these equations. Hence the state of stress beneath a strip coincides with the state of stress beneath a rectangle for $L \rightarrow \infty$ (cf. Fig. 21). For practical purposes, a rectangle with the side ratio $L/B > 4$ to 5 can already be regarded as a strip (Giroud, 1973, p. 275).

(b) No strip, i.e. infinitely long, load exists. It is, therefore, more convenient to employ a general procedure which the use of tables makes as simple as a plane problem.

As already mentioned, it is typical of plane problems that μ does not appear in any of the stress components, provided that the load is applied to the surface and the boundary conditions are specified in terms of displacements alone. This is the reason why in the case of plane strain, the stress σ_x in a saturated foundation soil experiences no changes in the course of consolidation (Lambe and Whitman, 1969, p. 489).

Approximate formulae, usually those for determining the variations of σ_z in a $z = \text{const.}$ plane, are also proposed sometimes. The oldest of such formulae is probably an empirical relation of Strohschneider (Strohschneider, 1912) also used by Terzaghi (Terzaghi, 1925, p. 222 et seq.). A special case of this formula is Boussinesq's relation describing σ_z. In a general case, however, the course of σ_z along the axis of the load is not smooth (Ohde, 1938). A simpler and more popular relation is that recommended by Kögler and Scheidig (1928, 1938) – see, for example, Florin (1959, p. 152 et seq.). Other empirical relations (proposed by Jáky, Bendel, Brinch Hansen) are described by Sherif (1973). All such relations lost their significance once tables and diagrams enabling a prompt determination of the exact value of stress had been set up.

2.3.2 *Internal loading of a space*

In contrast to a halfspace which is bounded by a plane at right angle to the z axis, a space (in the case being considered) is bounded in no direction.

Loading of an elastic (isotropic) space by a concentrated load applied inside the space, is the first of this group of problems which was ever solved. The solution was presented by Lord Kelvin (Thompson, 1848). Using the notation of Fig. 20, the stress components are defined by

$$\sigma_r = \frac{P}{8\pi(1-\mu)} \frac{r}{R^3} \left[\frac{3r^2}{R^2} - (1-2\mu) \right]$$

$$\sigma_\theta = -\frac{P}{8\pi} \frac{1-2\mu}{1-\mu} \frac{z}{R^3}$$

$$\sigma_z = \frac{P}{8\pi(1-\mu)} \frac{z}{R^3} \left[\frac{3z^2}{R^2} + (1-2\mu) \right]$$

$$\tau_{rz} = \frac{P}{8\pi(1-\mu)} \frac{r}{R^3} \left[\frac{3z^2}{R^2} + (1-2\mu) \right] \tag{54}$$

(because of symmetry, $\tau_{r\theta} = \tau_{z\theta} = 0$, σ_θ is the principal stress).

Unlike in eqns. (50) and (51), all stresses now depend on the mechanical parameters of the foundation soil (Poisson's ratio μ). This circumstance is typical of all solutions in which the load is not applied to the surface, or the boundary conditions are given in terms of the displacements (deformations).

Since the depth of foundation is always finite, this solution is only theoretically important, as the limiting case of load beneath the surface of a halfspace. As implied by eqns. (50) and (54) for $\mu = 0$ and 0.5, the stress σ_z along the axis of a load inside a space is only one third to one half of the stress σ_z under a load acting on the surface of a halfspace.

2.3.3 *Subsurface loading of a halfspace*

A solution for subsurface loading by a concentrated force was presented by Mindlin (1936). His formulae are fairly extensive, as for example, that for the stress σ_z at the point A under a vertical load (for notation, see Fig. 23), which is as follows:

$$\sigma_z = \frac{P}{8\pi(1-\mu)} \left[\frac{(1-2\mu)(z-h)}{R_1^3} - \frac{(1-2\mu)(z-h)}{R_2^3} + \right.$$

$$\left. + \frac{3(z-h)^3}{R_1^5} + \frac{3(3-4\mu)\,z(z+h)^2 - 3\,h(z+h)(5z-h)}{R_2^5} + \frac{30\,hz(z+h)^3}{R_2^7} \right] \tag{55}$$

Complete expressions, together with some results in table form are presented by Florin (1959) in the literature which is readily accessible. Integration of Mindlin's equations yields relations applicable to the case when the loading area is circular (Kézdi, 1952; the stress calculation was worked out in table form by Nishida, 1966) or rectangular (Škopek, 1960), under a uniform load. Practical calculations of stresses

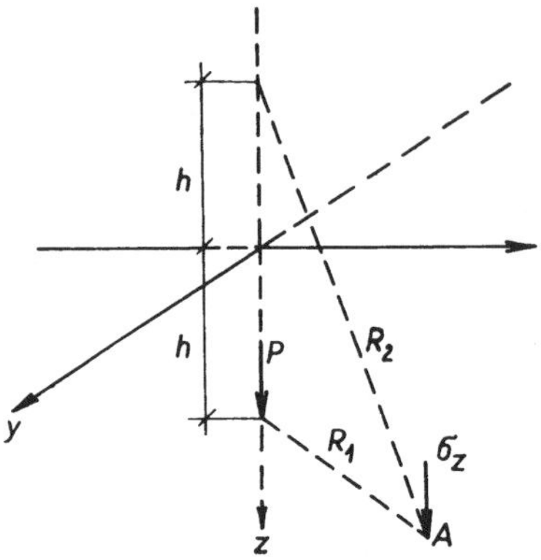

Fig. 23. Symbols used in the σ_z stress formula for a point load acting inside a halfspace.

beneath a rectangular loading area are conveniently carried out using tables contained in the thesis of Doroshkevich from 1959, an excerpt of which is given by Tsytovich (1963, p. 230)*). The stress σ_z beneath the centre and the corner of a rigid rectangular and beneath a circular foundation plate can be calculated after Butterfield and Banerjee (cf. Poulos and Davis, 1974, p. 180 et seq.).

The case of a load linearly distributed along a vertical line passing from the surface to a certain depth (which simulates the problem of skin friction of piles) can be solved by means of tables published by Geddes (1966).

A special case of the general solution is that of a line load (along the y axis) when plane strain is involved. This solution was presented by Melan (1932) before Mindlin's solution. Florin (1959) corrected it after Gorbunov-Posadov to eliminate an error which Melan allegedly committed. Stresses produced by a strip load can be obtained by integrating the line load.

Fig. 24 shows the distribution of the stress σ_z defined by eqn. (55) in planes parallel to the surface and lying above and below the point of action of the load. A part of the load is transferred by tension above the point of action, so that the compressive stress below this point turns out to be less than that in Boussinesq's solution. In many instances the subsoil transfers tension not at all or only partially, the main reason for this being the fact that the original subsoil in the vicinity of and above the foundation is replaced by backfill. The applicability of the results of Mindlin's solution is, therefore,

---

*) This author fails to state the value of μ for which the tables were prepared. According to private information, $\mu = 0{\cdot}35$ in this case.

controversial. Consequently, many authors give no, or only partial, consideration to the effect of the depth of foundation on the stress distribution. Jelinek (1951) with his approximate solution belongs to the last-named group.

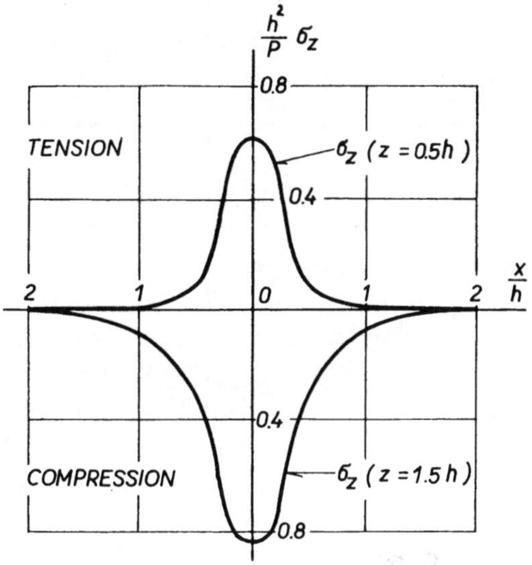

Fig. 24. Distribution of the σ_z stress in a horizontal plane above and below the point of application of a point load — according to Mindlin's solution (Florin, 1959).

2.4 Homogeneous anisotropic and linear continuum

The assumption of general anisotropy not only leads to very complicated solutions but makes it practically unfeasible to measure experimentally the twenty-one elastic constants of the subsoil which are needed in the calculation. Attempts of engineering importance concentrate, therefore, on a special case of anisotropy, the so-called cross anisotropy (also termed orthotropy with a plane of isotropy, hexagonal anisotropy or aeolotropy). In this case the mechanical behaviour of the subsoil is adequately described by two Young's moduli (E_v — in the vertical, and E_h in the horizontal direction), three Poisson's ratios (μ_1 — the effect of horizontal stress on horizontal strain, μ_2 — the effect of horizontal stress on vertical strain, μ_3 — the effect of vertical stress on horizontal strain; Poisson's ratios μ_2 and μ_3 are interdependent) and the shear modulus of elasticity G in the vertical plane (cf. Section 1.2.3). Next to the condition that the deformation moduli should be positive, the condition of positive strain energy imposes the following restrictions on these parameters (cf., for example, Poulos and Davis, 1974, p. 10):

$$1 - \mu_1 - 2\mu_2\mu_3 > 0$$

$$1 - \mu_1 > 0$$

$$1 + \mu_1 > 0 \tag{56}$$

These restrictions apply only to elastic materials and need not be fulfilled in real soils in which even the deformation moduli can in fact be negative (cf. Gibson, 1974).

Since a more general case (mentioned in Section 1.2.3) is less important for practice, the writer will subsequently concentrate on the discussion of a halfspace with the plane of isotropy parallel to the surface, i.e. horizontal (this assumption was also used in the definition of the above constants).

The idea of using the theory of a cross-anisotropic halfspace for the foundation soil, stems from the concept that soils have been formed by sedimentation. This concept logically implies that the mechanical properties of soils in the vertical direction should differ from those in the horizontal direction (all horizontal directions being equivalent to one another). The same assumption appears justifiable in preconsolidated soils where $K_0 > 1$. If $E_h = nE_v$, then according to experiments, $n \leqq 1$ for normally consolidated clays, $1 < n < 3$ for overconsolidated clays, and $n < 1$ for sands (Barden, 1963; Gerrard, 1975). For incompressible media $n \leqq 4$ (Gibson, 1974). On the basis of numerous published results, Gerrard (1975) states that anisotropy (characterized by the magnitude of n) is greater in rocks than in soils. According to Gerrard (1975), the large value of n of overconsolidated foundation soils is a consequence of horizontal (i.e. parallel to the stratification) fissures which are caused by comparatively high horizontal pressures induced by overconsolidation.

By triaxial tests of undisturbed samples of London clay (overconsolidated clay), Simons and Som (1969) have shown that the assumption of cross anisotropy is much more suitable for the description of experimental results (deformation of samples) than isotropic elasticity. What is remarkable is the fact that the crystals of hexagonal systems (Love, 1927) and sets of equal-diameter spheres in a hexagonal array (Gassman, 1951, 1953) have the properties of cross anisotropy.

Boussinesq's and Cerruti's solutions were extended by Michell (1900) to cover a cross-anisotropic halfspace. Wolf (1935) intending to apply it to foundation soils, presented it in a simplified form. Although Wolf employs three elastic parameters instead of five, which fact makes his results generally inacceptable (Barden, 1963)*), his solution is frequently quoted (Florin, 1959; Tsytovich, 1963).

In a general case when x, y, z are not the principal axes, the constitutive equations of a cross-anisotropic halfspace, eqns. (25), take the form

*) Despite this fact, solutions of stresses and strains in anisotropic media obtained by the use of three parameters (μ, E_h, E_v) continue to appear. Thus, for example, Milovic (1972) uses this method to solve the state of stress and deformations of a layer of finite thickness loaded by a stiff circular plate. The identity of Poisson's ratio is a special case of cross anisotropy which changes to isotropy for $n = 1$ and $2G = E/(1 + \mu)$. The lack of fulfilment of the last condition is the reason why the curve of K_0 for a 45° slope of the axis of isotropy (Fig. 9) has no common point of intersection with the other two, the condition $E_h = E_v$ being not sufficient for isotropy (see also Gibson, 1974).

$$\varepsilon_x = \frac{1}{nE_v}\sigma_x - \frac{\mu_1}{nE_v}\sigma_y - \frac{\mu_3}{E_v}\sigma_z\,, \quad \gamma_{xz} = \frac{\tau_{xz}}{G}$$

$$\varepsilon_y = \frac{1}{nE_v}\sigma_y - \frac{\mu_1}{nE_v}\sigma_x - \frac{\mu_3}{E_v}\sigma_z\,, \quad \gamma_{yz} = \frac{\tau_{yz}}{G}$$

$$\varepsilon_z = \frac{1}{E_v}\sigma_z - \frac{\mu_2}{nE_v}\sigma_x - \frac{\mu_2}{nE_v}\sigma_y\,, \quad \gamma_{xy} = \frac{2(1+\mu_1)}{nE_v}\tau_{xy} \qquad (57)$$

As a comparison of the above with eqns. (13) clearly shows, isotropy is a special case of cross anisotropy for $E_v = E_h = E$ (i.e. $n = 1$), $\mu_1 = \mu_2 = \mu_3 = \mu$, and $2G = E/(1 + \mu)$. Whenever the elastic parameters are thus reduced to two, Michell's formulae for the state of stress assume Boussinesq's form (cf. Barden, 1963).

From Michell's equation of the stress σ_z in a halfspace loaded on the surface by a concentrated load P (being too complicated, the equation is not reproduced here), Barden (1963) established the distribution of σ_z in a plane parallel to the surface (Fig. 25) for a range of possible combinations of μ_1, μ_2, μ_3 (the Poisson's

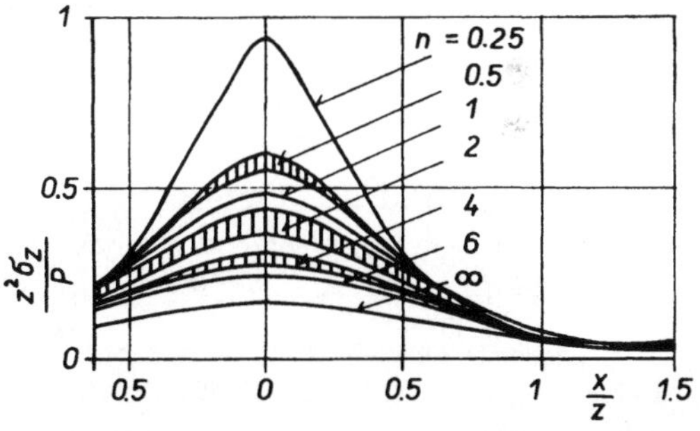

Fig. 25. Distribution of the σ_z stress in a plane parallel to the surface as a function of the ratio $E_h/E_v = n$ (Barden, 1963).

ratios are related by the equation $\mu_2 = n\mu_3$). As the figure shows, the stress concentration along the load axis (the z axis) decreases with increasing n. The ranges (done in hatching) for certain constant values of n indicate the way in which the corresponding distribution depends on the possible variability of Poisson's ratios. From the point of view of engineering practice these ranges may be regarded as so narrow that the effect of μ_1, μ_2 and μ_3 can be ignored and the type of the distribution of σ_z assumed to depend only on the ratio between the two Young's moduli, $n = E_h/E_v$ (the effect of Poisson's ratios on the horizontal stress cannot, however, be neglected). In contrast to Poisson's ratios, the ratio n can reliably be determined by experiments.

Proceeding from Barden's solution, Milović (1971) obtained σ_z beneath a corner of a rectangular area with different L/B ratios for several values of n. His results are in qualitative agreement with those shown in Fig. 25.

Barden's solution is only of limited validity because it is based on an implicit assumption restricting the generality of the shear modulus G (Dooley, 1964). It seems, however, that Barden's assumption concerning the shear modulus being independent of the rotation of the coordinate axes can be acceptable – at least in certain cases (in London clay whose $n = 2$, for example, to an accuracy of 5% – Gibson, 1974). A general solution was presented by Koning (1957); some of the results of this author are shown in Fig. 26 which is in qualitative agreement with Fig. 25.

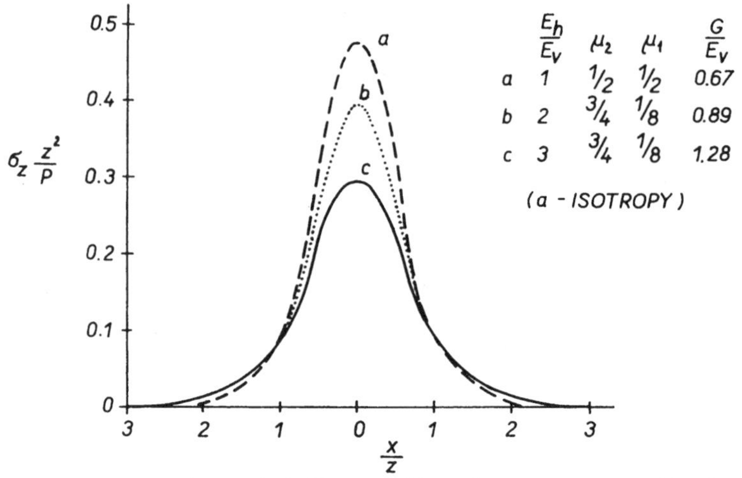

Fig. 26. Stress σ_z in a cross-anisotropic halfspace subjected to a concentrated load acting on the surface – according to Koning's (1957) general solution.

Practical applications of the general theory of cross-anisotropic halfspace in soil mechanics are made possible by studies of Gerrard and Harrison (1970a,b), particularly by the first of these which contains tables necessary for the calculation in the case of a loaded circular area. The second study is devoted to the solution of a strip load in a more general case of anisotropy – an orthorhombic halfspace (nine parameters of the foundation soil, three mutually perpendicular planes of elastic symmetry). The load may be vertical as well as shear, uniform and linear (Fig. 22), in the case of a circle, also shear radial and torsional, and by combinations of these loads (since the principle of superposition applies, cf. eqn. (57)) one can simulate a general load.

Application of the theory of cross-anisotropic halfspace calls for the knowledge of five deformation parameters which can be determined by tests in a triaxial apparatus using samples with vertical (E_v, μ_1), horizontal (E_h, μ_3) and oblique (G) axes of isotropy – Silveira and Souto Silveira (1973), and Gibson (1974) who puts forward the reasons why a 45° obliqueness is the most convenient of all. The first four parameters can be found directly, by a test in the apparatus, using a sample with a general state of stress (Morgan and Gerrard, 1973). In the case of an axi-symmetrical stress, eqns. (57) reduce to eqns. (25) which imply the relation for K_0. Once K_0 is known, it is enough to perform two triaxial tests (isotropic compression and standard compression test) to define the practically important ratio between the axial and the radial strain of the sample (Simons and Som, 1969).

A special case of cross anisotropy, i.e. a foundation soil which does not deform horizontally, was tackled by Westergaard (1938). It can be shown that Westergaard's model of the foundation soil is equivalent to a cross-anisotropic medium with three independent parameters. Since $\varepsilon_x = \varepsilon_y = 0$, the normal stresses are only a function of ε_z (Poulos and Davis, 1974, p. 192), and $\sigma_x = \sigma_y = K_0\sigma_z$. Practical applications of Westergaard's solutions are enabled by a diagram which can be used for the determination of the stress σ_z beneath the corners of a loaded rectangle (Taylor, 1948), and by Newmark's diagram of this solution (Sowers, 1962; Bowles, 1968). Compared with Figs. 25 and 26, the curve of propagation of the stress σ_z in a plane parallel to the surface (Taylor, 1948, p. 253) shows that Westergaard's solution of σ_z approximately corresponds to the distribution curve for $n \sim 3$ depicted in these figures.

2.5 Non-homogeneous linear continuum

Most important for engineering practice and best refined theoretically is a solution relating to two types of non-homogeneous halfspaces. The first is a halfspace whose Young's modulus varies continuously with depth. Such a halfspace is a continuously non-homogeneous continuum. The other is a stratified halfspace, the subsoil being homogeneous within the range of each stratum, i.e. its Young's modulus is constant but changes suddenly, discontinuously on the boundaries between the strata. Hence a stratified foundation soil (the surfaces of the layers are assumed to be parallel with the ground surface) is termed a discontinuously non-homogeneous continuum.

Unless otherwise specified, both types are assumed to be isotropic.

2.5.1 Continuous non-homogeneity

As an analysis of eqns. (50) has shown, in an incompressible medium ($\mu = 0.5$) the propagation of stresses in a halfspace is radial, the point of action of the force P acting as a source. All normal stresses are either zero or compressions. For $\mu < 0.5$ the propagation of stresses ceases to be radial and tensile stresses appear in the subsoil (Fröhlich, 1934, p. 14 et seq.). If the subsoil (for example, sand) is incapable of transmitting these tensions, the stresses must propagate radially. This reason, together with the simplicity of solution and the fact that this case of stress propagation is a special case of the exact Boussinesq's solution (for an incompressible halfspace) led Fröhlich (1934) to the elaboration of a complete stress theory on the basis of the law of radial stress propagation. For a halfspace loaded on the surface by a concentrated vertical load (Boussinesq's problem), Fröhlich (1934, p. 24) defines the stress σ_R (Fig. 20) by

$$\sigma_R = \frac{\nu P}{2\pi R^2} \cos^{\nu-2} \varphi \tag{58}$$

where $\cos \varphi = z/R$ (Fig. 20). The quantity ν is the statically indeterminate so-called concentration factor. For $\nu = 3$, one obtains

$$\sigma_R = \frac{3P}{2\pi R^2} \cos \varphi \tag{59}$$

which is Boussinesq's relation for an incompressible isotropic halfspace. For a uniaxial state of stress according to eqn. (58), one can deduce the following relations for σ_z and for the other non-zero components of the stress tensor:

$$\begin{aligned}
\sigma_z &= \frac{\nu P}{2\pi R^2} \frac{z^\nu}{R^\nu} \\
\sigma_r &= \frac{\nu P}{2\pi R^2} \frac{r^2}{R^2} \frac{z^{\nu-2}}{R^{\nu-2}} \\
\tau_{rz} &= \frac{\nu P}{2\pi R^2} \frac{z^{\nu-1}}{R^{\nu-1}} \frac{r}{R}
\end{aligned} \tag{60}$$

In a somewhat different form the first of eqns. (60) was already proposed in 1932 by J. H. Griffith who made use of Goldbeck's measurements (cf. Fröhlich, 1934, p. 178) and by Ivanov (1926) according to Soviet authors (Tsytovich, 1963, p. 214; Florin, 1959, p. 157; Klein, 1956). Analogous formulae for a vertical surface load, a linear load, a load acting on a circular area, etc. were derived by Holl (1940).

As Fröhlich (1934, p. 90 et seq.), starting from the condition of minimum deformation work, has already shown, the case of $\nu = 3$ applies to an elastic isotropic halfspace (E = const.), $\nu = 4$ to a halfspace whose Young's modulus increases linearly with depth. If it is, generally,

$$E = E_0 \left(\frac{z}{z_0}\right)^m \tag{61}$$

(z_0 – a length unit, sometimes replaced by the width B or the diameter D of the foundation), then for $\nu = 3$, $m = 0$, and for $\nu = 4$, $m = 1$. In this interpretation the concentration factor ceases to be a statically indeterminate quantity which can be chosen empirically. Eqn. (58) becomes an exact solution for a continuously non-homogeneous halfspace. For $m \neq 0$, E_0 is equal to the magnitude of E for $z = z_0$. Eqn. (61) can also be understood to mean that Young's modulus E is a function of the vertical effective geostatic pressure according to the relation

$$E = E_0 \delta_0^{-m} \left(\frac{\gamma_0 z}{\gamma_w z_0}\right)^m \tag{62}$$

where $\delta_0 = \gamma_0/\gamma_w$.

The first author to formulate Fröhlich's interpretation generally was Ohde (1938) who assumed that the variation of Young's modulus obeys the equation

$$E = E_0 \left(\frac{z + d_0}{z_0} \right)^m \tag{63}$$

and stated that the conditions of equilibrium, Hooke's law and the geometrical conditions of deformation require that the following relation between the quantities ν, m and μ (Poisson's ratio)

$$\frac{1}{\mu} = \nu - 1 = 2 + m \tag{64}$$

should be satisfied in the comparatively simple case of $d_0 = 0$ (for $d_0 \neq 0$ this solution is only approximate). In the special case of isotropy, $m = 0$, and this fact requires that there should be $\mu = 0.5$ and $\nu = 3$. It is clearly seen from eqn. (64) that one can leave out the concentration factor and write an equation which only relates μ to m. A solution more general than that of Ohde (1939) was published by Hruban (1945). If a halfspace is continuously non-homogeneous in the sense of eqn. (61)*), two ways of radial propagation of stresses are possible (the following equations are somewhat modified Hruban's relations):

$$\text{a)} \quad \mu = \frac{1}{m + 2} \quad \text{(Ohde's case)} \tag{65}$$

$$\text{b)} \quad \mu = \frac{1}{m + 1} \tag{66}$$

A special case of these equations for $m = 1$ ($\mu = 0.5$ and 0.33) was treated by Borowicka (1943). The magnitude of σ_R is then defined by (in contrast to Hruban, the + sign is affixed to pressures)

$$\text{in the case of a)} \quad \sigma_R = \frac{m + 3}{2\pi} \frac{P \cos^{m+1} \varphi}{R^2} \tag{67}$$

$$\text{in the case of b)} \quad \sigma_R = \frac{m + 2}{2\pi} \frac{P \cos^{m} \varphi}{R^2} \tag{68}$$

These equations will change to eqn. (59) on the condition that $m = 0$ (eqn. (67)) or $m = 1$ (eqn. (68)); both values of m lead to $\mu = 0.5$ but the halfspace is homogeneous only in the first case. Similar relations (with the additional use of the concentration factor) are proposed by Klein (1956) without reference to Hruban's work.

Thus Hruban's equation (68) admits the possibility of existence of an incompressible non-homogeneous halfspace with a homogeneity of the type described by eqn.

*) Hruban (1945) also presented solutions of other, more complicated cases of non-homogeneity, for example, of radial non-homogeneity.

(61), which was later studied by Gibson (1967). Some of the many publications which have appeared in sequel of Gibson's work, deal exclusively with the settlement (for example, the papers by Biarez and Boulon, 1971; Brown and Gibson, 1972, 1973; Awojobi, 1975; Simons and Rodrigues, 1975) and will, therefore, be discussed in Section 3.2.

Gibson (1967) considers the following two types of non-homogeneity of the subsoil:

$$G = G_0 + a\,\frac{z}{z_0} \tag{69}$$

and

$$G = G_0\,\frac{d}{d - z} \tag{70}$$

where a is a parameter, and d – the depth at which $G \to \infty$. Since the medium is isotropic, the shear modulus of elasticity, G, can be replaced by $E/2(1 + \mu)$ in eqns. (69) and (70). A special case of both relations ($a = 0$ or $d \to \infty$) is homogeneity. Since the hyperbolic relation (70) has not been used except by Awojobi (1975) in an analysis of deformations, only eqn. (69) will be considered in the discussion that follows.

Using Gibson's terminology, one speaks of so-called "partial heterogeneity" if $G_0 \neq 0$, and of so-called "complete (full) heterogeneity" if $G_0 = 0$ in eqn. (69).

If $G_0 = 0$, eqn. (69) is identical with eqn. (61) for $m = 1$; if $G_0 \neq 0$, it is identical with eqn. (63) for $m = 1$. Using eqns. (65) and (66) one can write eqn. (61) generally in the form

$$E = E_0 \left(\frac{z}{z_0}\right)^{(1-2\mu)/\mu} \tag{71}$$

and

$$E = E_0 \left(\frac{z}{z_0}\right)^{(1-\mu)/\mu} \tag{72}$$

For complete non-homogeneity, eqn. (69) becomes

$$G = a\,\frac{z}{z_0} \tag{73}$$

Eqn. (73) is clearly a special case of eqns. (71) and (72) for $\mu = \frac{1}{3}$ in the former, and $\mu = \frac{1}{2}$ in the latter equation, and $a = \frac{3}{8}E_0$ and $\frac{1}{3}E_0$. Hence, when eqn. (73) is applied, the stresses in a halfspace propagate radially only for $\mu = \frac{1}{3}$ and $\frac{1}{2}$; otherwise, tensions develop in the subsoil and it should be ascertained whether or not the subsoil is capable of transmitting them.

Gibson's solution of 1967 makes use of eqn. (73) and $\mu = \frac{1}{2}$, i.e. it is applicable to an incompressible halfspace. As shown by a comparison of eqn. (68) with eqn. (59), in this case the state of stress in an incompressible homogeneous halfspace does not differ from that in a fully non-homogeneous halfspace (although the state of strain in the two halfspaces is completely different, as also acknowledged by Gibson, 1967).

Gibson and Sills (1971) demonstrated that this agreement between the states of stress does not apply to the general case of $\mu \neq \frac{1}{2}$ (except for $\mu = \frac{1}{3}$, tensions arise in the subsoil under such conditions, as indicated): the stress σ_z is somewhat higher and the stress σ_x (the case under study was that of plane strain) is substantially lower in a non-homogeneous compressible halfspace than in a homogeneous halfspace.

A solution of the state of stress beneath a uniformly loaded strip and a circular area for partial non-homogeneity defined by eqn. (69) and a general value of μ was presented by Awojobi and Gibson (1973). For similar properties of the subsoil, a case of practical importance, i.e. that of a perfectly rigid circular foundation, was solved numerically (by means of the finite element method) by Carrier and Christian (1973). They found that the effect of roughness of the foundation base is important only in the case of low non-homogeneity and $\mu < 0.3$, and that it is, therefore, enough – for practical purposes – to regard the foundation base as smooth. As their solution shows, the stress σ_z as well as the stress σ_r depend on both the degree of the foundation soil non-homogeneity and the magnitude of Poisson's ratio μ. As μ tends to 0.5, the first dependence gradually vanishes. This is at considerable variance with the situation in a homogeneous halfspace in which σ_z is independent of μ (the first of eqns. (50)). In the case of complete non-homogeneity the dependence becomes apparent from eqns. (67) and (68) following the substitution of eqns. (65) and (66). One can then conclude that for the stress σ_z produced in the axis of a concentrated load P, eqns. (67) and (68) merge into a single equation, viz.

$$\sigma_z = \frac{(1/\mu) + 1}{2\pi} \frac{P}{z^2} \tag{74}$$

which is identical with eqn. (59) ($\varphi = 0$, $R = z$) or with the first of eqns. (50) in the case of $\mu = \frac{1}{2}$. If $\mu \neq \frac{1}{2}$, it is different from that for σ_z in a homogeneous halfspace.

Gibson (1974) also published an analysis of the case of an incompressible non-homogeneous cross-anisotropic halfspace. The condition of incompressibility reduces the number of the foundation soil parameters from five to three – E_h, E_v and G. If each of these parameters increases linearly with depth according to eqn. (73), the results are analogous to those obtained for a non-homogeneous isotropic halfspace.

The properties of continuously non-homogeneous subsoils in the case of plane strain were also studied by some authors. For a more general case of non-homogeneity described by eqn. (63), one may propose to use the solution for a substitute half-plane with non-homogeneity defined by eqn. (61), and to determine stresses at a distance $(z + d_0)$ from its surface. Since the boundary condition is not satisfied, the difference in stresses on the boundary of the actual half-plane can be regarded as an additional load whose effect is subtracted by the help of the substitute half-plane.

For subsoils, $m = 0.5$ is frequently taken for the most typical value of the exponent m in eqn. (61) (Hruban, 1945; Lambe and Whitman, 1969, p. 224) – see also Section 3.2; in keeping with eqn. (65), this value of m leads to $\mu = 0.4$. However, judging by the results of numerous laboratory tests of undisturbed samples (Skempton and Henkel, 1957; Ward et al., 1965; Paál, 1971; Wroth, 1971; etc.) as well as by those of field plate loading tests (Burland et al., 1973), eqn. (63) for $m = 1$ and eqn. (69) seem to be more suitable. From the laws of uniaxial compressibility of soils, one obtains $m = 0.5$ for sands, and $m = 1$ for clays (Feda 1976c; cf. Section 3. 12).

2.5.2 Discontinuous (layered) non-homogeneity

A critical review of the solutions of stresses in continua with layered non-homogeneity was presented by Kafka (1964). The most common and of greatest significance for practice, are the solutions relating to layered continua of the following two types: with alternating layers (in the simplest case, a two-layer subsoil) of elastic but mutually different properties, and with an elastic layer underlain by an incompressible base. In either case the layers are assumed to be parallel to the surface.

The first type of continuum is of special interest to transport engineering. Classical among the studies devoted to it are the publications of Burmister (1943, 1945, 1956) and their practical value is further enhanced by tables set up for two layers (Gerrard, 1969; Gerrard and Morgan, 1969 – treating a circle; Giroud, 1973 – treating a strip) as well as for three layers (Jones, 1962, and Giroud, 1972 – both treating a circle). The solutions proposed by Burmister for a two- and a three-layer halfspace in closed form, can also be obtained by means of the finite element method (Duncan et al., 1968).

Provided that the properties of the layers are not so markedly different as in the case of a compressible soil on a rock base, or a solid to hard layer on a soft layer, and provided that the thickness of the upper layer is not at least 1/4 to 1/2 of the width of the loading surface (Florin, 1959), this model of halfspace is not particularly important for the calculation of settlement.

Approximate solutions in which the friction between layers is ignored (Marguerre, 1933) considerably magnify the effect of non-homogeneity (Rappoport, 1948). If the difference between the deformation moduli is small, σ_z increases by up to several tens of percent in the case of a smooth contact (cf. Fox, 1948a, the solution of a two-layer system for $\mu = 0.5$.) As a more detailed analysis of the effect of gradually decreasing contact roughness indicates, the stress experiences a change only at a very smooth to perfectly smooth contact and remains virtually unvaried for a perfectly rough to medium rough contact (Novotný, 1975, for $\mu = 0.35$). For μ tending to zero the effect of roughness of the contact between layers becomes negligible.

A general solution relating to a large number of alternating elastic layers was published by Kafka (1964). His solution applicable to deeper layers is a fitting supplement to Burmister's theory covering the state of stress in surface layers. It can be shown that a subsoil formed by layers of two alternating homogeneous, isotropic materials is equivalent to a homogeneous cross-anisotropic medium, provided that the overall thickness of the layered subsoil is small compared to the controlling dimensions of the problem under consideration (Poulos and Davis, 1974, p. 191). According to Steinbrenner (1934), the stress in a layered subsoil can in many cases be replaced by the stress in a homogeneous subsoil without too great an error, provided that the deformation modulus increases with depth (Giroud, 1969, 1970b; 1972, p. 11; 1975; Poulos and Davis, 1974, p. 163), and provided that the contact between the individual layers is not smooth (Giroud, 1972, p. 11). This possibility is implied by the solution obtained in Section 2.5.1 for a non-homogeneous incompressible halfspace, which shows that the state of stress in a non-homogeneous halfspace is the same as that in a homogeneous halfspace.

Filon (1904) was the first to solve the effect of an incompressible base on the state of stress in an overlying elastic layer as a two-dimensional problem, Biot (1936) the first to solve it as a three-dimensional problem. Compared with a homogeneous halfspace, the stress σ_z generally always concentrates along the load axis (as can also be deduced by analogy with a continuously non-homogeneous halfspace at $m > 0$ according to eqn. (67)). The extent of this concentration depends to a degree on the magnitude of Poisson's ratio (this effect is frequently regarded as unsubstantial – Florin, 1959; Poulos, 1967; Giroud et al., 1970; Poulos and Davis, 1974, p. 113, and others) but very markedly on the circumstance whether a zero (smooth contact) or a non-zero (rough contact) shear stress is introduced as the boundary condition on the contact with the incompressible base. Just as in a two-layer elastic subsoil, the concentration of the stress σ_z is considerably lower in the latter case (cf. for example, the solution for a rectangular load and a smooth – Sovinc, 1961 – or a rough –

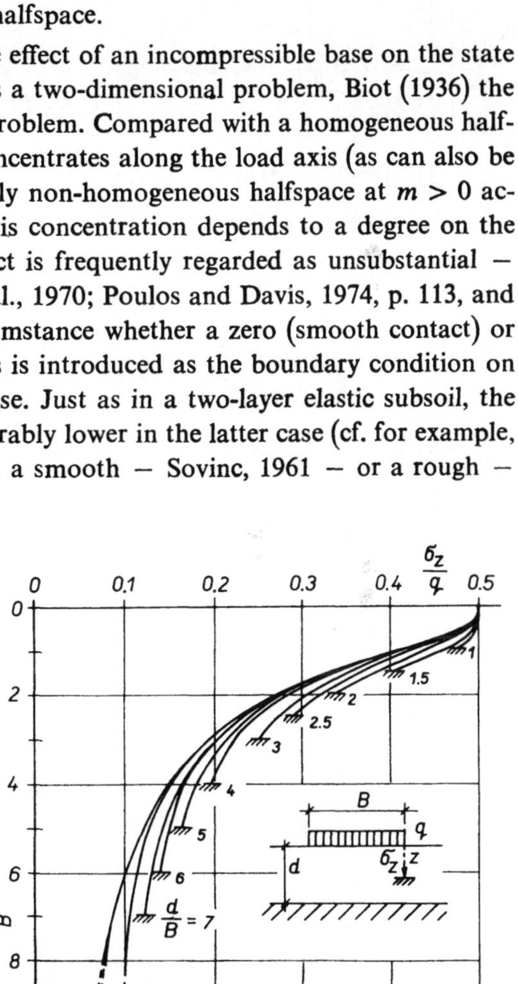

Fig. 27. Effect of an incompressible and rough base on concentration of σ_z stress beneath the edge of a foundation strip (σ_z at the centre is obtained by simple superposition) — according to Giroud et al. (1970) for $\mu = 0.3$.

Milovic and Tournier, 1971 – surface of the incompressible base). The σ_z stress concentration for the case of a rough base can approximately be evaluated by a graphical procedure (Giroud et al., 1970). The effect of an incompressible and rough base at various depths on the stress concentration is shown in Fig. 27.

2.6 Non-linear continuum

A medium is physically non-linear if its deformation parameters depend on the state of stress. Three actual cases of this dependence have been studied:

(a) The shear modulus of elasticity depends non-linearly on the second invariant of the stress deviator (or: on the stress intensity, octahedric shear stress) – Hruban (1958).

(b) The (normal) modulus of elasticity depends linearly on the first invariant of the stress tensor (or: on the mean principal stress, octahedric normal stress) – Huang (1968).

(c) The (normal) modulus of elasticity depends non-linearly on the mean principal stress σ_{oct} – Biarez and Boulon (1971).

The first type of halfspace was studied by Hruban (1958).

If the strain intensity ε_i is defined by the equation

$$\varepsilon_i = \frac{1}{\sqrt{2}} \frac{1}{1+\mu} [(\varepsilon_1 - \varepsilon_2)^2 + (\varepsilon_2 - \varepsilon_3)^2 + (\varepsilon_3 - \varepsilon_1)^2]^{1/2} \tag{75}$$

and the stress intensity σ_i by the equation

$$\sigma_i = \frac{1}{\sqrt{2}} [(\sigma_1 - \sigma_2)^2 + (\sigma_2 - \sigma_3)^2 + (\sigma_3 - \sigma_1)^2]^{1/2} \tag{76}$$

(where σ_1, σ_2, σ_3, and ε_1, ε_2, ε_3 are the principal stresses and strains), then, according to Hruban (1958), for a physically non-linear halfspace

$$\varepsilon_i = \frac{\sigma_i}{G_i} \tag{77}$$

where G_i is the deformation modulus dependent on σ_i. A special, practically important case of eqn. (77) is the parabolic deformation law

$$\varepsilon_i = \left(\frac{\sigma_i}{G}\right)^k \tag{78}$$

where G is the constant deformation modulus, i.e. the deformation modulus of a linearly elastic body ($k = 1$; k is a parameter). Assuming a radial stress propagation (with tension in the halfspace excluded) a procedure similar to that used in Section 2.5.1 enables the following two relations to be derived for a homogeneous non-linear halfspace:

$$\text{a)}\quad 2k = \frac{1}{\mu} + 1 \tag{79}$$

$$\text{b)}\quad 2k = \frac{1}{\mu} \tag{80}$$

Either eqn. (79) or eqn. (80) must be fulfilled in order that the solution might satisfy the conditions of equilibrium and compatibility. For $k = 1.5$ to 3, Hruban (1958) gives the magnitude of the stress σ_R from which one can deduce (the case under consideration being that of radial stress propagation) the remaining components of the stress tensor. In keeping with eqn. (80) a linear halfspace is a special case of a non-linear halfspace for $\mu = 0.5$ (then $k = 1$). Some relations applicable to a continuously non-homogeneous, non-linear halfspace can also be established.

Huang (1968) starts from the dependence of Young's modulus E on the first invariant of the stress tensor, proposed by Barkan, which can be written in the following form:

$$E = E_0(1 + 3\alpha_0\sigma_{oct}) \tag{81}$$

where $\alpha_0 = \text{const.}$, and from the assumption that the medium is incompressible ($\mu = 0.5$). He also takes into account the effect of the geostatic state of stress so that Young's modulus reaches its minimum at medium depths and increases near the surface (due to the effect of the external load) as well as at greater depths (due to the effect of the geostatic stress).

Huang (1968) performs the calculation by dividing the subsoil into seven layers (for a uniform circular surface load); he assumes $E = \text{const.}$ within the range of each of the layers and obtains its magnitude by successive integration. An analysis of the accuracy of the calculation shows this procedure to be acceptable.

Biarez and Boulon (1971) assume E to be dependent in the following way:

$$E = \alpha_0\sigma_{oct}^k \tag{82}$$

where $0 < k < 1$, and describe the effect of k on σ_z along the axis of a circular, uniformly loaded area (for $\mu = 0.3$) and on the settlement of the centre and the edges of the area. The effect is important only so far as the magnitude of the settlement is concerned.

As the results of all three authors show, the propagation of vertical pressures in a physically non-linear halfspace does not substantially differ from that in a linear halfspace (Hruban, 1958, p. 35; according to Huang, 1968, the difference is 30% max.). Hruban's solution yields a lower σ_z stress concentration along the load axis in the non-linear than in the linear halfspace. Huang (1968) finds the same to be true at smaller depths, and the reverse to apply to depth greater than about $3D$ (D – the diameter of a circular loading area). This conclusion cannot, however, be regarded as general. Both solutions start from the special condition of physical non-linearity (78) or (81) instead of using a general non-linear relation between the stress tensor and the strain tensor. The introduction of such a general non-linearity condition is enabled by a numerical solution by means of the finite element method (cf. Section 3.8). Wibel (1971) employed this method to solve the state of stress in a non-linear sandy subsoil loaded by a circular foundation. He obtained a higher concentration of the vertical stress along the foundation axis than in the case of a linear halfspace. Qualitatively similar results were also obtained by Biarez and Boulon (1971).

The fact that the concentration of the vertical stress along the foundation axis is different in a non-linear than in a linear halfspace, can be put to use in practice: this solution offers a further possibility, besides the continuously non-homogeneous halfspace and the cross-anisotropic halfspace, of describing different concentrations of the stress σ_z. The application of the results of the solution is made considerably more difficult by the circumstance that the principle of superposition does not apply to non-linear halfspace. A comparison of eqns. (81), (82) with eqn. (62) points out close connection between the theories of non-homogeneous and non-linear halfspaces, which can be utilized in practice. As demonstrated by Biarez and Boulon (1971), one solution can be replaced by the other accurately enough whenever the course of $E(z)$ is approximated with reliability.

A remarkable progress in practical applications of the non-linear stress-strain relations was brought about by the use of numerical methods, which has been made possible by automatic computers (they were already employed by Huang, 1968, and Biarez and Boulon, 1971). Thanks to its versatility, the finite element method is regarded as the most useful of all the numerical procedures so far devised.

The non-linear behaviour of real soils in an oedometer is frequently described by the equation (Lambe and Whitman, 1969, p. 157)

$$\sigma_z = a(\varepsilon_z)^b \tag{83}$$

where a, b are parameters; very often, particularly for cohesionless soils, $b = 2$. For a perfectly arranged set of elastic spheres, $b = 3$. The difference in the value of b in a real medium is attributed to the effect of displacements among the particles and their regrouping.*)

2.7 Unified conception of non-homogeneity, anisotropy and non-linearity

As eqns. (67) and (68) imply, the greater m is, the higher is the stress concentration along the load axis. From this point of view, the case is analogous to that of a cross-

*) Under hydrostatic stress, this effect is suppressed and as a result, the volume deformation is usually only elastic (Thurairajah and Roscoe, 1965).

anisotropic medium (cf. Figs. 25 and 26) – the increase of m is analogous to the decrease of n $(= E_h/E_v)$. This fact was first pointed out by Barden (1963). According to eqn. (64), one can replace the exponent m by Fröhlich's concentration factor and seek the analogy of the latter to the ratio of the deformation moduli n. This is the way in which the analogy was formulated by Milović (1971).

Biarez and Boulon (1971), on the other hand, draw attention to the analogy between the solution of a non-homogeneous halfspace (the non-homogeneity being defined by eqn. (69)) and that of a non-linear halfspace (the non-linearity being defined by eqn. (82)). They claim that in the majority of cases, the non-linear solution can be changed to non-homogeneous with an accuracy of 4% to 5%.

The analogy between non-homogeneity and anisotropy was systematically studied by Gol'dstein (1971), who takes into consideration only Ohde's relations (64), (65) and (67) (attributing them to Holl, 1940, and Klein, 1956) and compares the magnitude of σ_z along the axis of a vertical force acting on the surface of the halfspace (Fig. 20). The writer adopts the same procedure. In the respective equations $\varphi = 0$ and $R = z$.

From the first of eqns. (50) one obtains

$$\sigma_z = \frac{3}{2\pi}\frac{P}{z^2} \tag{84}$$

which can be written as

$$\sigma_z = K\frac{P}{z^2} \tag{85}$$

where $K = 0.477$ for an isotropic, linear and homogeneous halfspace. For a non-homogeneous halfspace, the magnitude of the parameter K can be evaluated from relations (61) or (73) using eqns. (67) and (68). The magnitude of K (the ordinate) as a function of n which characterizes the measure of subsoil anisotropy can be taken (with possible graphical interpolation and for average values of μ) from the diagram in Fig. 25. Hruban's (1958) solution for the state of stress in a non-linear halfspace defined by eqn. (78) can similarly be used to determine the magnitude of K for the condition (80) (eqn. (79) yields $K = 1/\pi =$ const. for all k).

The results of these calculations can be set up in a table (Table 1) arranged in the following way:

The effect of non-homogeneity is as defined by eqn. (67). Values shown in parentheses in columns 4 and 6 correspond to the effect of non-homogeneity described by eqn. (68). The difference between the respective values corresponds to the effect of Poisson's ratio according to eqns. (65) and (66). As these equations indicate, the solution of m in column 2, just as that using eqn. (68), assumes that $-1 \leqq \mu \leqq 1/2$ (cf. Gibson, 1974). The values of k given in columns 5 and 6 were graphically extrapolated when necessary (Hruban, 1958, confines his solution to $k = 1$ to 3), and serve for reference only. By eqn. (80) one obtains $-\frac{1}{2} \leqq k \leqq \infty$ for $-1 \leqq \mu \leqq \frac{1}{2}$.

Table 1

K	m (eqns. (67), (61))	$n\,(=E_h/E_v)$ (Fig. 25)		k (eqns. (78), (80))	
1	2	3	4	5	6
0.96	3	0.25	(0.30)	$-\frac{1}{2}$	$(-\frac{1}{4})$
0.80	2	0.30	(0.45)	$-\frac{1}{4}$	$(+\frac{1}{4})$
0.64	1	0.45	(1)	$+\frac{1}{4}$	(1)
0.56	$\frac{1}{2}$	0.55	(2)	0.6	(2)
0.48	0	1	(3.5)	1	(5)
0.40	$-\frac{1}{2}$	2	(6.5)	2	—
0.32	-1	3.5	(∞)	5	—
0.24	$-1\frac{1}{2}$	6.5	—	—	—
0.16	-2	∞	—		
0.08	$-2\frac{1}{2}$	—	—		
0	-3				

It is clearly seen from Table 1 that in a number of cases one can achieve equal concentration of σ_z in a non-homogeneous, in an anisotropic as well as in a non-linear halfspace. Thus, for example, one obtains the same σ_z for linear non-homogeneity ($m = 1$), cross-anisotropy with $n = 0.45$ and parabolic non-linearity with $k = \frac{1}{4}$, etc. Hence, an experimentally ascertained stress concentration in a subsoil can theoretically be explained in three alternative ways: by a change of Young's modulus with depth (continuously non-homogeneous halfspace), by a difference between Young's moduli in the vertical and the horizontal direction (cross-anisotropic halfspace), or by the dependence of Young's modulus on the state of stress in the subsoil.

Although these three different mathematical models are nearly equivalent from the point of view of the σ_z stress prediction, they are unlike in other aspects. Compared with the remaining two, a cross-anisotropic halfspace is clearly at a disadvantage so far as the difficulties associated with the measurement of physico-mechanical parameters of the subsoil are concerned. The invalidity of the principle of superposition is a shortcoming of the non-linear halfspace model. Hence, one can regard as the most suitable the theory of non-homogeneous halfspace which, as the above analysis suggests, is capable of approximately modelling also the effect of anisotropy and non-linearity of the subsoil on its state of stress.

2.8 Experimental results of stress measurements

2.8.1 Effect of the method of measurement

A comparison of experimental and theoretical results must be preceded by a critical interpretation of the measured values, for a method of measurement might happen to affect the results much more markedly than do the mechanical properties of the medium being examined. Whenever this is the case, the experimental results cease to be significant enough for testing theoretical conclusions concerning stress propagation.

Of primary importance is the measuring device proper—usually an earth pressure cell. Two types of such cells are in use: a strain-gauge type in which one measures the deflection of the diaphragm of the cell and uses this value to determine the applied load by means of a calibration curve, and a liquid type in which one measures directly the stress of the liquid filling the cell. More generally distributed stresses are better measured with the former than with the latter type.

As is well known, the results of measurement strongly depend on the relation between the deformation parameters of the cell and the medium being measured (Taylor, 1947; Monfore, 1950; Peattie and Sparrow, 1954; Tsitovich and Baranov,

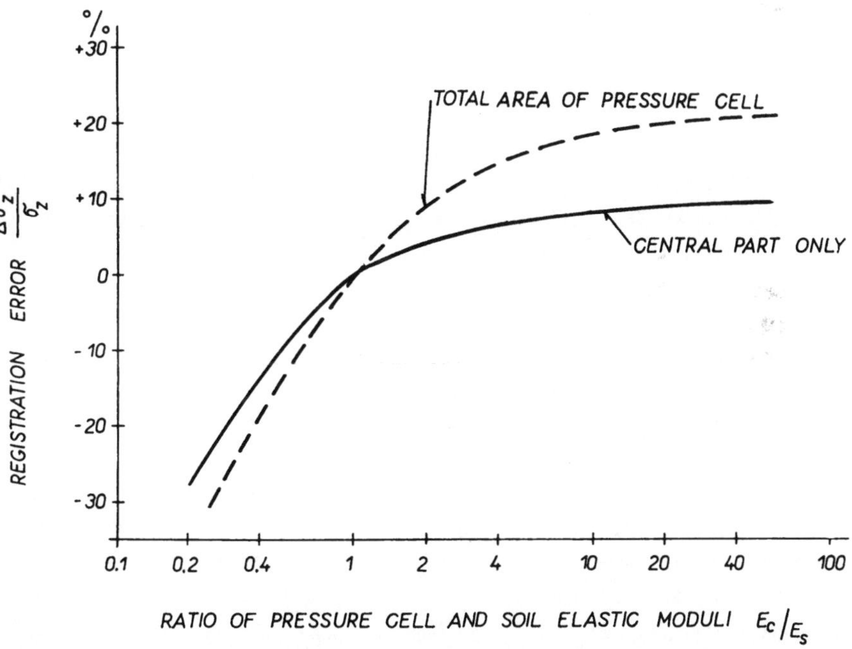

Fig. 28. Relation between the earth pressure cell error (for a cell with $v/D = 0.2$; v, D — cell height and diameter) and the ratio between Young's moduli of the cell and the subsoil (Whiffin and Morris, 1962).

1961; Tsytovich, 1963; Thomas and Ward, 1969; Brown, 1973; Marchetti, 1975; and others). The ratio between the difference of the measured and the actual stress, $(\sigma_z^{exp} - \sigma_z) = \Delta\sigma_z^{exp}$ and the actual stress σ_z, the so-called earth pressure cell (registration) error depends on the ratio between Young's moduli of the cell, and of the subsoil; a theoretical curve of this dependence is shown in Fig. 28 (after Whiffin and Morris (1962) – similar results were also obtained by Prange (1965) and others). This effect is frequently termed the flexibility factor and expressed as a ratio between the stiffnesses of the cell and of the subsoil; next to the deformation parameters of the subsoil it also includes the geometric parameters of the cell (cf. Brown, 1973; Marchetti, 1975). In order that the contact stress of the cell might be distributed with reasonable uniformity, the gauge section of the cell should not be larger than about 1/2 of its area (Peattie and Sparrow, 1954).

Referring to Fig. 28, it has been found convenient to use fairly stiff cells (Tsitovich and Baranov, 1961, recommend that the deformation modulus of an earth pressure cell should be one order of magnitude higher than the deformation modulus of the subsoil, i.e. 1 to 1.5×10^3 MPa) because the error of such cells is constant, independent of the deformation modulus of the medium.

If the earth pressure cell is low enough (Tsitovich and Baranov, 1961, recommend that v/D should be less than 0.1, where v – the height, and D – the diameter of the cell; as early as 1944, the U.S. Waterways Experimental Station specified that the so-called aspect ratio, v/D, should be less than 0.2), then the earth pressure cell error can be defined by Taylor's relation

$$\frac{\Delta\sigma_z^{exp}}{\sigma_z} = a\,\frac{v}{D} \tag{86}$$

where a is a factor which can be established theoretically or experimentally and whose magnitude depends on the kind of soil involved (cf., for example, Tsytovich, 1963 – sand: $a = 1.2$; stiff clay: $a = 0.6$ to 1.5; plastic clays: $a = 0.2$) and on other parameters (the deformation modulus of the soil and the cell, and Poisson's ratio). In keeping with eqn. (86), the stress concentrates in the earth pressure cell during measurement, and the measured stress is, therefore, larger than the actual stress (see also Fig. 28). The cell diameter depends on the gradient of the measured stress field and should be at least fifty times larger than the diameter of grains of the subsoil in which measurements are made.

If $E_c/E_s > 10$ and $v/D < 0.1$, then, according to Tsitovich and Baranov (1961), the total error of measurement does not exceed 12% to 14%. However, the accuracy of measurement also depends on the magnitude of the recorded pressure (for example, Kravtsov, 1970, puts the maximum error at 7% for $0.15 < \sigma_z < 0.4$ MPa, and up to 35% for $0.02 < \sigma_z < 0.05$ MPa).

The effect of the stiffness of earth pressure cells was already acknowledged by Strohschneider (1912). He refers to Forchheimer's tests during which the rubber balls used to measure the pressure deformed excessively and, as a result, indicated only 5%

of the actual stress. Strohschneider used cells equipped with a device to annul their volume deformations. To eliminate arching ("soft" earth pressure cells transmit their loads to less yielding surroundings and the cell records lower stresses) Kögler and Scheidig (1934, p. 77) have used – supposedly with considerable success – a large number of cells (which were comparatively soft) placed next to one another. Goldbeck's stiff cells in which the external pressure was compensated pneumatically (Terzaghi, 1925, p. 381) were evidently too high. In strain-gauge earth pressure cells, the requirement imposed on the accuracy of measurement of the diaphragm deflection leads to a certain, though usually acceptable compromise concerning the stiffness of such cells (Brown, 1973). The most suitable from this point of view are strain-gauge cells with piezoelectric crystals, and liquid-filled cells. Both can also be built with a very low aspect ratio but perfect de-aeration must be ensured for the latter type (Marchetti, 1975). The principle of strain gauging and filling with liquid are both used in a cell developed for the U.S. Federal Highway Administration, which has a piezoresistive gauge and is filled with oil (Brown, 1973). Morgan and Gerard (1968) used miniature (1″ dia.) pressure cells while Prange (1965, 1971) worked with somewhat larger cells – v = 15 mm, D = 40 mm – which transmitted the recorded values by wireless. A special type of gauges, so-called aerostatic gauges, was proposed by Pokrovskiy and Fedorov (1936, 1953) for field measurements and centrifuge model tests.

Errors also occur as a result of improper calibration of the earth pressure cells. Calibration must be carried out in the measured medium and for the recorded state of stress. As Petrianin (1970) and Petrianin et al. (1972) have shown, under equal hydro-

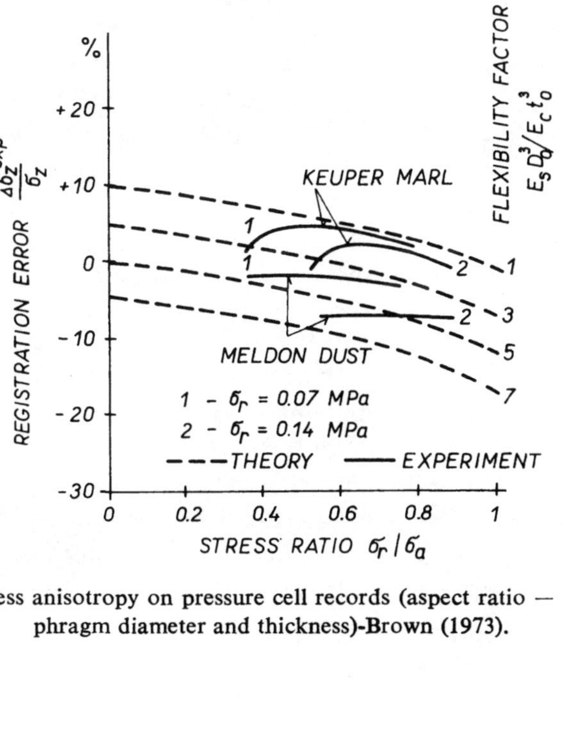

Fig. 29. Effect of stress anisotropy on pressure cell records (aspect ratio — 0.17; D_0, t_0 — diaphragm diameter and thickness)-Brown (1973).

static pressure a pressure cell yields different values for sand and glycerine, and its function is significantly affected by stress anisotropy and the stress path. The theoretical as well as the experimental effect of stress anisotropy established by Brown (1973) is shown in Fig. 29. In the case under consideration this effect altered the measured stresses by 6% max.

Other inaccuracies arise owing to boundary conditions which the theoretical solution fails to take into account. Fröhlich (1934) points out the effect of shear stresses in the base of a loading plate which, in his opinion, raise the stress σ_z (cf. Section 2.3.1), and the effect of the magnitude of the load which is apt to produce local plastic regions in the subsoil.

Some distortion of the results of laboratory tests is brought about by the limited dimensions of the test box. On the basis of Melan's solution (Melan, 1918), Ohde (1938) states that the effect of a stiff base is apt to raise the value of the concentration factor by as much as 1.5.

Whenever stresses are measured in an undisturbed foundation soil, disturbances in the inherent structure of the medium during the process of inserting the earth pressure cells in the soil clearly become extraordinarily important. An extreme case of this are the results of Kenney (1967) (extreme in the sense that the tested soil was a sensitive clay) obtained during the measurements of the coefficient K_0: the differences due to the effect of these disturbances were of an order of up to 100%.

In the tests made by Kravtsov (1970) precautions were taken to ensure that the volume of the subsoil disturbed by the insertion of the earth pressure cells and the gauge marks should not exceed 5% of the volume of the active zone (subsoil region whose deformation produces settlements). Tsytovich (1969) describes stress measurements performed in silt with a fairly high salt content (bottom of a shallow lake). According to this author, the natural structure (cementing by salts) of this soil was completely renewed five to seven days after having been disturbed.

It is clearly seen from the above analysis that the measurement of stress propagation in foundation soils is accompanied by considerable difficulties. Under optimal conditions one might expect an accuracy of about 15% to 20%. Significantly refined results may be obtained by the elimination of systematic errors of measurement on the basis of the requirement that the volume (area) of the pattern of the measured stress should be equal to the external load. Kandaurov (1966, p. 280) asserts that the fact that the earlier measurements of which he made a systematic analysis, generally agree with the results obtained by Baranov using more modern (Plantema's) earth pressure cells, can be explained by just this correction.

The effect of measurement inaccuracies will clearly in many instances cause the experimental interval of measurement reliability to be broader than the interval of comparable theoretical solutions and thus make the experiment lose its decisive function.

2.8.2 *Results of measurements*

The state of stress is measured phenomenologically (the diameter of the earth pressure cell is much larger than that of the soil grains), the discrete nature of the subsoil coming into play only when measuring the neutral stress. But even in this case one must measure the total stress to evaluate the effective stress. If it is only stresses that are being measured, it is frequently impossible to decide which of the mathematical halfspace models satisfies the real subsoil. Thus, for example, the σ_z stress concentration ascertained along the foundation axis can be interpreted with the aid of a model of a non-homogeneous, an anisotropic as well as a non-linear halfspace (cf. Section 2.7). The situation can be made unambiguous only by complex measurements by means of which one finds not only stresses but also displacements at various points of the subsoil.

The results of measurements can be divided into three groups:

(a) direct measurements which immediately yield the state of stress (total stresses) of the quasi-continuum,

(b) indirect measurements which make it possible to estimate this state of stress only after additional intermediary relationships (for example, eqn. (46) when measuring the neutral stress) have been introduced,

(c) complex measurements which supply data not only on the state of stress but on the displacements of various points of the subsoil as well.

2.8.2.1 *Direct measurements*

A critical review and evaluation of earlier measurements of the stress σ_z in a surface loaded artificial sandy subsoil was presented by Fröhlich (1934). The review includes the measurements made by Steiner and Kick (Prague, 1879), Strohschneider (Graz, 1909–1911, cf. Strohschneider, 1912), University of Illinois (1910–1913), Pennsylvania State College (1913–1914), Goldbeck (Washington, 1917), Kögler and Scheidig (Freiberg, 1925–1927) and Hugi (Zürich, 1927). Since Fröhlich also cites the literature in which the tests are described, the respective references are not listed here.

Fröhlich arrives at the conclusion that the experimental stress distribution is concentrated more towards the load axis than it should according to Boussinesq's equations: the concentration factor $\nu = 6$ for small, heavily loaded areas (where the effect of plastic regions in the subsoil manifests itself), and $\nu = 3$ to 4 for large areas (lightly loaded and therefore with a negligible effect of plastic regions) or for loads acting at some depth. The results reported by the Pennsylvania and Illinois Universities are in good agreement with the results of Kravtsov for soaked-through loess and of Baranov for sand (cf. Kravtsov, 1970). Ohde (1938) adds to Fröhlich's conclusion that, after deducting the effect of the stiff bottom of the test box, the results of measurement yield $\nu = 3$ to 4 on the average.

Just as Fröhlich (1934), Kandaurov (1966) quotes nearly all earlier measurements and supplements them by the measurements of Soviet authors (Laletin, 1930; Pokrovskiy and Fedorov, 1932–1933; Baranov, 1955–1960; Bolshtianskiy, 1962). He finds an excellent agreement to exist between the statistical theory and the measured results (the quantity η in eqn. (49) was, however, not measured independently but was chosen so as to obtain the best possible agreement). It should be mentioned at this point that the difference between the variations of the σ_z stress with depth according to the linear (elasticity) theory and those according to Fröhlich's or statistical theories is greater by far than the difference between the two last-named theories.

Pokrovskiy and Fedorov (1936, 1953) describe the results of measurements of stresses in a sand fill artificially prepared under field conditions (in a 6.5 × 3.5 × 3 m foundation pit) using aerostatic gauges, and in a laboratory centrifuge. They measured both the σ_z and the σ_x stresses. Their results also show a somewhat greater concentration of the stress σ_z along the axis of the loading plate (having an area of 5000 cm^2 and loaded with 0.25 MPa).

Bolshtianskiy (1962) presents the results of measurements of compacted sand gravel with a clayey base, which he believes to be in agreement with the statistical theory of stress distribution evolved by Kandaurov (cf. Section 2.2.2).

Turnbull et al. (1961) report about the results of extensive tests performed by the U.S. Waterways Experimental Station. All three normal stresses, σ_x, σ_y and σ_z, and also two normal stresses on surfaces inclined at an angle of 45° to the vertical were measured during these tests. The artificially prepared subsoil was clayey silt and sand, the loading surface had a diameter of 28.5" (72.5 cm). On the basis of an analysis the authors claim that the results correspond well with an isotropic and homogeneous elastic halfspace for $\mu = 0.5$ (clayey silty subsoil) and $\mu = 0.3$ (sand). They explain the slightly higher stress concentration in sand by the effect of the pressure cells. Lapkin's (1975) measurements of the vertical stress in a model of a cross-anisotropic medium (layers of loess and lime-cement mortar) confirm the stress concentration to be lower than in an isotropic medium.

Tsytovich (1969) published results of measurements made on a supposedly undisturbed foundation soil (a lake bottom formed by salty silt) loaded by a plate ($D = 112$ cm). The accuracy of measurement was about 12%. Up to a depth of $z = D$, the stress σ_z beneath the centre and the edge of the plate was in good correspondence with that in a model of linear halfspace (deviations of 10% to 12%); at greater depths it was higher by 30% to 40%.

2.8.2.2 Indirect measurements

In indirect measurements the theoretical stress field is estimated indirectly from the measured quantity using an additional relationship. Since the procedure simultaneously involves testing of this auxiliary relationship, it is sometimes impossible to estimate the variations of the theoretical stress field unequivocally.

The most valuable of indirect measurements are the measurements of neutral stresses in saturated clayey foundation soils under increasing loads. The calculations are made using the theory of isotropic elastic halfspace and Skempton's equations (46) or (47). The reason why the results are so useful is the fact that they generally relate to field tests of subsoils which are not disturbed except by the insertion of the piezometric gauge. Provided that no local plastic regions arise, the measurements show an excellent agreement with theory under increasing load, as well as under slightly decreasing load (Lambe and Whitman, 1969, pp. 401 and 404; Hoëg et al., 1969). The agreement is somewhat worse for greater unloading (Lambe and Whitman, 1969, p. 404) and, possibly, under cyclic load variations; in the latter case, however, at least a qualitative agreement can still be noted (Gibson and Marsland, 1961).

Tsytovich (1963, p. 307) describes an indirect method which consists of deducing the stress from the familiar relation between the pressure and moisture content of clays by determining the moisture content at various points of the subsoil after the termination of the loading test.

The measurement of settlement can be said to be another indirect method provided that the procedure used combines the theoretical stress field with an experimental stress-strain relation of the subsoil (i.e. the compression curve of undisturbed samples, in situ loading test, etc.). Because of the predominant effect of the deformation properties of the subsoil, this is the least reliable of all the indirect methods.

More important are the results of laboratory studies of strains at various points of the subsoil. Aichhorn made such tests in 1931 using wooden balls, Gerber in 1932 using lead balls whose position was determined by means of X-rays (cf. Tschebotarioff, 1951, pp. 208 and 200). Gerber's method was employed by Bergfelt (1956) for an undisturbed sample of clay, and by Burland (1969) for an artificially prepared clayey subsoil. Bergfelt believes that the disturbance produced by the insertion of lead markers (balls and cubes) was negligible. The results of measurements of vertical displacements along a vertical passing through the centre of a square loading plate agree very well with Boussinesq's theory in Steinbrenner's application (Bergfelt, 1956 – Fig. 15). Similar field measurements are discussed in Section 3.10.

Less straightforward still are the indirect methods in which no comparison is made of either stresses or strains but which directly test the suitability of the chosen mathematical model of halfspace. The results of measurements of mechanical properties prove full justification of models of cross-anisotropic halfspace (with different deformation moduli in the vertical and the horizontal direction), of non-homogeneous halfspace (increase of the deformation modulus with depth, different deformation moduli of various layers of the subsoil) and of non-linear halfspace (non-linear stress-strain relation). These will be discussed in more detail in Section 3.9.

2.8.2.3 *Complex measurements*

Examples of such measurements were published by Lomize and Kravtsov (1969), Kravtsov (1970) and Lomize et al. (1972). The measured values of vertical and horizontal displacements at various points of the subsoil are approximated by continuous functions which are used to calculate ε_x, ε_y and ε_z. Their sum – the volume strain – is checked by direct measurement. Normal stresses in three perpendicular and one oblique directions (Kravtsov, 1970) or possibly, shear stresses (Lomize et al., 1972) are measured at the same time.

The tests were performed in situ, in loess soaked in advance, and loaded by a circular foundation ($D = 1.6$ m), as well as in laboratory, in sand under the conditions of plane strain (a model foundation with $B = 41.6$ cm).

The results relating to the sandy subsoil (Lomize et al., 1972) showed a fairly good agreement of the stress σ_z with the theory of linear halfspace (the error was less than 20%, the concentration of the stress σ_z along the foundation axis was slightly higher); the results relating to the remaining stress components were of equal order of magnitude except for a region immediately beneath the footing bottom (of a width and depth of about 0.75 B) where the measured differences amounted up to 200%. In loess (Lomize and Kravtsov, 1969; Kravtsov, 1970) the differences in σ_z reached as much as 190% and in σ_r as much as 620%; however, the theoretical solution assumed a flexible plate although the test plate was stiff. In this case, also, the concentration of the stress σ_z was higher than that corresponding to a linear halfspace.

More significant than the direct comparison between the theoretical and experimental states of stress was an analysis of the experimental stress-strain relations. This analysis revealed that the condition of coaxiality and similarity of the stress tensor and the strain tensor is not satisfied, and above all, that the volume strain depends to a considerable degree on the second invariant of the stress deviator (the effect of dilatancy). This seems to suggest that none of the constitutive relations of an elastic continuum is physically appropriate, and that, largely, concurrence between theoretical and experimental states of stress has merely a formal character and defies physical interpretation.

In the tests of sand, the relations between stresses and strains in the subsoil corresponded to those obtained in measurements of laboratory samples loaded homogeneously (Lomize et al., 1972). A way towards a physically well grounded solution of the state of stress in the subsoil should, therefore, proceed from more general (including, for example, the effect of dilatancy) experimentally verified constitutive equations specific to particulate matters.

A report on unusually instrumented complex measurements of stresses and strains in a model sandy subsoil subjected in laboratory to a uniform load applied to a 10″ (2 5.4 cm) dia. area was published by Gerrard and Morgan (1972). In conformity with the identified anisotropy and non-linearity of the sand, the theoretical results relating to a cross-anisotropic and to a non-linear halfspace agreed on the whole with the experimental values, although they displayed larger deviations in the case of shear stresses. Their agreement with the results for an isotropic and a linear halfspace was substantially worse.

2.9 Conclusion

The following two criteria can provide a base for judging the suitability of the various methods for calculating the stress fields in the subsoil: a) how easily a method can be applied to calculations, and b) how well it approaches reality.

(a) In keeping with the first criterion, one should estimate how much a particular method is theoretically worked out to meet the various boundary conditions encountered in practice, and how readily it can be employed in engineering calculations. In this respect the theory of elastic homogeneous isotropic halfspace indisputably excels all others. Its great advantage lies in the fact that no experimental data are needed for the calculation of the stress σ_z since this is frequently completely independent of the mechanical properties of the foundation soil. This holds approximately true even in the case when local plastic regions arise in the subsoil (Sections 2.2.1 and 3.4).

Quite well applicable is the theory of non-homogeneous halfspace combined with Fröhlich's concentration factors. This calculation procedure can in fact be employed universally, i.e. even in the case of the non-linear halfspace theory (which, just as the statistical theory, is otherwise unsuitable for calculations) and in the case of the cross-anisotropic halfspace theory (which can be applied using studies of Gerrard and Harrison, 1970a,b and tables and diagrams prepared for Westergaard's solution). The magnitude of a suitable concentration factor is derived either from the variations of Young's modulus, $E(z)$, using eqns. (81), (82) and (61) and (64), or from the expected concentration of the stress σ_z according to Fig. 25 and the first of eqns. (60). Table 1 can also be used, and the non-homogeneity determined from the value of Poisson's ratio and eqns. (65) and (66). Carrier and Christian (1973) propose to determine the subsoil non-homogeneity on the basis of the settlement of loading plates of different diameters subjected to equal unit loads.

(b) The second and principal criterion should be a physically well grounded agreement with experimental results. Such results are, however, scarce and mostly insignificant from practical standpoint because they were obtained in tests of artificial fills of sand. Moreover, the accuracy of measurement is fairly low. All this notwithstanding, the results seem to indicate that the concentration of the stress σ_z along the load axis is usually higher than that calculated by means of the theory of linear isotropic and homogeneous halfspace. Although more substantial differences are hard to prove, they may be expected to occur in other components of the stress tensor rather than in the stress σ_z. The results of complex measurements (Section 2.8.2.3), though still few and far between, pose the all-important question whether a possible agreement between theory and experiment is not frequently of merely a formal character, i.e. a matter of mathematical analogy. If the answer were in the affirmative, the possibility of extrapolation would become a considerable problem and a physically justified agreement would have to be sought in the essentially differently conceived theories based on qualitatively different constitutive relations.

In view of the limited accuracy, impossible unambiguous physical interpretation (Table 1; in the case of incompressibility, the state of stress in a homogeneous halfspace is the same as in a completely non-homogeneous halfspace, etc.) and difficulties associated with stress measurements which were discussed above, the decisive answer must be expected from settlement calculations and their comparison with reality. As will be shown in Chapter 3, many procedures different from the elasticity theory actually find application in settlement calculations.

Taking into consideration the existing results of measurements as well as the theory which predicts similar or even identical states of stress for physically entirely different halfspaces, many experts are inclined to use the simplest theory of elastic homogeneous and isotropic halfspace (particularly for the calculation of σ_z; the deviations of the other stress components may be substantially larger) in combination with sound engineering judgement (Lambe and Whitman, 1969, p. 105 – the calculated stress may be inaccurate to $\pm 25\%$ or more; Tsytovich, 1963, pp. 307 and 308). Speaking more precisely, Florin (1959, pp. 81 and 82) adds that it is incorrect to ponder globally whether or not this theory is suitable for soils but that one should decide in which region (depending on the stress level, etc.) it represents a reasonable approximation.

Since the foundation soil is not elastic, it is more to the point to classify this model as a model of a linearly deforming halfspace (what is essential is the validity of Hooke's law) subject to the condition that it only applies to monotonically increasing loads. One does not inquire in which cases the foundation soil is incapable of taking up the tensions which this model predicts (cf. Section 2.5.1). Hence, from a physical point of view, this model is a priori only approximate for foundation soils even if the subsoil were to deform linearly.

As both the direct and indirect experimental results show, there exists a real possibility of accepting the theory of non-homogeneous and cross-anisotropic halfspace. The physical properties of these halfspaces correspond much better to those of actual foundation soils. The price which must be paid for this better physical match are experiments needed to define the parameters n (and possibly, other deformation parameters) and m. Direct experimental evidence derived from in situ stress measurements, has, however, been missing so far. Since the field measurements of the stress σ_z are extremely difficult (it is probably hardly possible to prevent the foundation soil from being disturbed to a high degree) and theoretical differences are expected to be small, one could test these theories by measuring the stress σ_x using a method similar to that described by Kenney (1967). This would, of course, require advanced information on whether the accuracy of measurement is sufficient for distinguishing between the various theories.

The availability of automatic computers of large enough capacity fundamentally alters the criteria of suitability of the various methods of calculation. The mathematical aspects lose their meaning, and the effort to introduce realistic constitutive relations becomes of primary concern. The most versatile in this respect is the finite element method which is capable of taking into account non-linearity, non-homogeneity and anisotropy of a medium, different types of geostatic stresses (which can also be calculated – cf. Malina, 1969) as well as complicated boundary conditions.

Computers can be used in two ways:

(a) To solve classical problems which have so far defied solution because of complicated boundary conditions (for example, unloading of the foundation soil by an excavation) or a complicated algorithm of the solution (for example, Burmister's method for solving a layered halfspace). Equations defining stresses produced by differently distributed loads have lost their meaning – the solution is carried out by summing up the effects of concentrated forces that substitute for the actual loads; one can exactly compute the extent of the plastic regions in the subsoil, apply the statistical theory to stress calculations (Kandaurov et al., 1969), etc.; thus, for example, Harrison et al. (1972) published a computer programme for stresses and strains in a homogeneous as well as in a layered, isotropic, cross-anisotropic and orthotropic subsoil subjected to strip and circular loads.

(b) To introduce realistic constitutive relations, particularly non-linearity. This is mainly important for the calculations of strains. The success of this procedure depends on how well one manages to determine experimentally the constitutive relations of a medium, on the variability of mechanical parameters with the stress path and stress level, etc.

In the writer's opinion, computer calculations, far from making the existing methods less useful, supplement and enlarge them, mainly so far as special cases (large, economically profitable problems; requirements imposed on the accuracy by important and extensive structures and projects) and research activities are concerned (see also Section 3.10).

From an engineering point of view, stress calculations usually represent a preparatory stage of settlement calculations whose reliability is a matter of primary interest. If various stress theories are used, the difference in stresses which the application of this procedure brings about is generally regarded as secondary in its effect on the magnitude of the settlement. This conclusion can in part be based on the finding that in plane problems the stress field produced by a surface load does not vary with the mechanical properties of the subsoil; in three-dimensional problems, this frequently holds true at least for the stress component σ_z*). As a point of first importance one

*) What matters is not whether the halfspace is more or less compressible but whether it is homogeneous or non-homogeneous, isotropic or anisotropic, etc.; however, even then — as mentioned above — the difference in stresses is largely unimportant.

mainly regards the choice of the mechanical parameters of the subsoil (i.e. particularly those which characterize compressibility) – cf. Hobbs (1974), further, the determination of the correct magnitude of the load, the effect of unloading of the foundation soil by excavation, etc. (see Section 3.13).

Thus the problem of stress propagation does not seem to play a particular role from the point of view of settlement calculations. But it certainly is of primary interest in the calculations of the neutral stress, for in some cases the magnitude of the neutral stress frequently decides the stability of the whole structure (for example, reservoirs and embankments on soft saturated clays, earth dams on a subsoil of low bearing capacity, etc.). It can hardly be said that the problem of stresses in a multiphase space has definitely been solved by the introduction of the model of a linearly deforming halfspace where the effect of non-linearity is described by the coefficient of pore pressure. Lambe's results (cf. Section 2.2.1) clearly show that for the time being this procedure cannot be regarded as final.

Chapter 3

METHODS FOR CALCULATING THE FINAL SETTLEMENT

3.1 Introduction

Engineering structures settle for many reasons: owing to the effect of additional loading of the subsoil by a structure, lowering of the groundwater level, diverse forms of ground surface sinking (mining, sliding, subsidence, underground erosion), etc. (Sowers, 1962; Terzaghi and Peck, 1967). This chapter will deal only with settlements produced by additional loading of the subsoil by a structure.

Methods used for calculating the final settlement form four qualitatively distinct groups of solution (Fig. 30). The first group includes mathematically exact methods

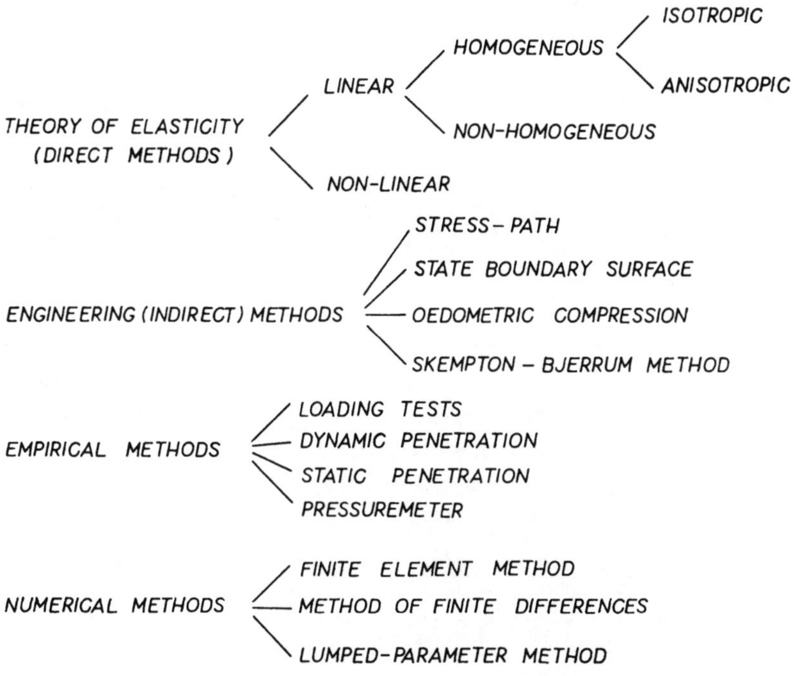

Fig. 30. Review of methods for calculating the final settlement.

of the elasticity theory which satisfy the equilibrium conditions, the constitutive equations and the boundary conditions prescribed. The most refined of these methods, the theory of elastic, homogeneous, isotropic and linear halfspace, is mathematically exact but its constitutive equations do not usually correspond to real foundation soils; hence the results obtained by its means are at considerable variance with reality. Thus, for example, the settlement of the surface of a homogeneous halfspace loaded (vertically or horizontally) by a strip is infinite, just as the settlement of a non-homogeneous halfspace loaded by a strip or a circular area, provided that the halfspace is not incompressible and the value of m in eqn. (61) is not less than unity (Awojobi and Gibson, 1973; Klein, 1956)*). Moreover, the selected mathematical model of the halfspace makes it impossible to predict time variations of the settlement; it is mainly employed in calculations of instantaneous (immediate, initial) settlement.

The second group encompasses engineering methods which are most widely used for settlement calculations. This group of methods makes use of the elasticity theory only for stress calculations; the stress-strain relations needed for the subsequent phase of the calculation, are derived from experiments. This is the reason why these methods, in contrast to the former, direct methods, are called indirect (Giroud, 1972, p. 10). The stress field is supposed to be invariant to a considerable degree with respect to the constitutive relations as well as to the mechanical properties of the subsoil**). The methods of this group readily introduce the non-linearity of the stress-strain relation of a real subsoil, and this fact is clearly the main reason why the calculated and the actual settlements are generally in good agreement.

The third group contains empirical methods. Empirical or semiempirical formulae are used to extrapolate for actual foundations the results of loading or complex mechanical tests (for example, penetration tests), particularly of cohesionless foundation soils.

The methods of the fourth group are highly demanding numerically and therefore require the application of automatic computers. The first and foremost of these methods is the finite element method. They are characterized by the introduction of real non-linear stress-strain relations (and possibly, of anisotropy and non-homogeneity), and the accuracy of the solution from the mathematical standpoint is limited only by the capacity of the computer. Although the solution is usually quasi-elastic, some of the methods are capable of also describing the effect of local plastic regions. As mentioned in the Introduction, these methods do not represent – from a factual viewpoint – an independent group, for they can also be applied to the elasticity theory

---

*) This is a result of a physically incorrect assumption concerning an infinite depth of the compressible subsoil. This error may be corrected by assuming a finite depth of the compressible subsoil, the so-called active zone, which will be discussed later (cf., for example, Tsytovich, 1963, p. 564; Schmertmann, 1970; Sherif, 1973, p. 42; Garga and Quin, 1974).

**) According to the foregoing, this assumption is evidently best satisfied for the stress σ_z. The reasons for this are discussed in Section 3.2.

(as was done, for example, by Giroud, 1972, p. 13; Giroud et al., 1970). Their classification as a separate group is, however, convenient methodologically.

Solutions obtained by the application of the methods of the second group frequently put together several components to make up the magnitude of the final settlements. Such components are: instantaneous (immediate) settlement, settlement produced by the effect of primary (filtration) consolidation, settlement due to secondary consolidation (creep), and settlement caused by plastic (inelastic) pushing of the subsoil from beneath the foundation to the sides (Seed, 1965).

Since the time variations of settlements are not included in subsequent considerations, a discussion of secondary consolidation (creep) is not presented. For the same reason, only the settlement following the termination of primary consolidation ("consolidation settlement" hereafter) rather than its time variations is the subject of interest in this chapter. The last problem, which is not considered here, is the effect of plastic pushing-out of the subsoil. At the time when the oedometric method, which excludes lateral deformations from the settlement calculations, was the only one available of the second group, some attempts were made to describe this effect quantitatively (Kögler and Scheidig, 1938, p. 119; Fröhlich, 1934, p. 109; etc.) However, more refined methods pay heed to lateral deformations of the subsoil, except in the case when the load approaches the limit load and extensive plastic regions are formed as a result. From the point of view of settlement calculations, this case is, however, insignificant. As some analyses (Hoëg et al., 1968; Hoëg et al., 1969) suggest, the vertical stress is to a great degree independent of plastic deformations and mechanical properties of soils, and this fact speaks in favour of the usual procedure of calculating the vertical stress from the elasticity theory, as practised by the methods of the second group.

In an analysis of the methods of calculation belonging to the second group, the final settlement is therefore understood to mean the sum of the instantaneous and the consolidation settlements.

Before proceeding to the description of the various methods of final settlement calculations, it seems useful to add three notes (on the effect of foundation stiffness, on the unloading by excavation, and on the depth of foundation) which apply to a number of these methods.

(a) The distribution of the contact stress varies with varying stiffness of foundations; this fact causes the base of foundation to deform during the settlement to a greater or lesser degree. In the extreme case of perfect stiffness, the base remains plane. If it is not plane, the settlement can be calculated for various points of the foundation area, for example, by the superposition of rectangles. Since most stress calculations were derived for a perfectly flexible foundation (usually a uniformly loaded rectangular or circular area) while actual foundations are stiff, it is advantageous to calculate the settlement for the so-called characteristic point of the foundation base. This is a point whose settlement is virtually independent of the distribution of the contact stress. In a rectangular or a square foundation and strip, it lies at a distance of 0.577 $B/2$ or 0.577 $L/2$ from the axis of symmetry, in a circular area at a distance of 0.707 $D/2$ from its centre (Marivoet, 1948).

(b) In the course of subsoil unloading by an excavation to a depth d intended for the founding of a structure, the bottom of the foundation pit experiences a heave*). The total load q acting on the subsoil thus unloaded by the structure, can be divided into two parts. In the first part – up to a load equal to the original geostatic pressure at the level of the base of foundation, $q = \gamma_0 d$ – the subsoil is repeatedly compressed and the problem is analogous to that of an overconsolidated subsoil. The settlement is approximately equal to the previous heaving of the bottom of the foundation pit. In the second part – for loads greater than the original geostatic pressure, $q > \gamma_0 d$ – the soil is compressed like a virgin, normally consolidated subsoil. If the subsoil is overconsolidated, the first phase of the load takes place up to the overconsolidation load. Hence the calculation of the settlement in the first phase of loading uses different deformation parameters of the subsoil than does the second phase.

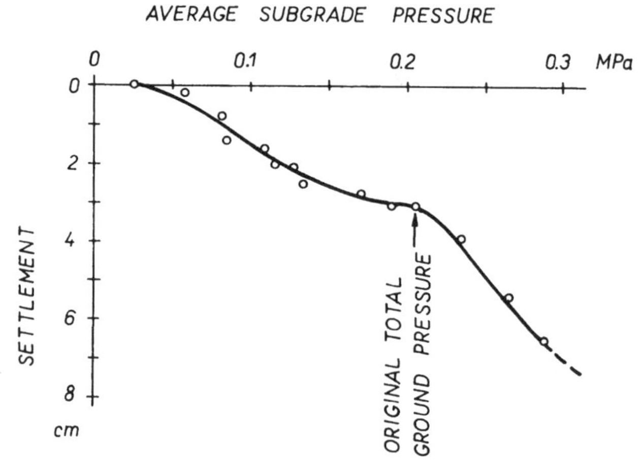

Fig. 31. Load-settlement diagram of a raft foundation — nuclear power station founded on sandy subsoil (Dunn, 1974a).

In the simpler case of normally consolidated subsoil, the load-settlement curve, like the compressibility curve in an oedometer, should display a distinct break which marks the magnitude of the original geostatic load $\gamma_0 d$.

An illustrative example of the effect of subsoil unloading by an excavation is shown in Fig. 31. In this case the subsoil is 2.5 times more compressible for $q > \gamma_0 d$. When the loading by a structure is equal to the original ($q = \gamma_0 d$), i.e. when the net increase

*) Heaving of the bottom of a foundation pit can be reduced, sometimes very substantially, by preventing the bottom from being completely unloaded. This is achieved by piles installed beneath the base of the foundation before beginning the excavation (Tornstensson, 1973; Butler, 1974).

in foundation pressure is zero (floating foundation), the settlement is practically equal to the previous heave of the bottom of the foundation pit. This conclusion was verified by a number of measurements (see, e.g. Simons, 1974; Ueshita et al., 1974). The effect of subsoil unloading by an excavation is, therefore, customarily introduced in the settlement calculations by adopting the procedure outlined above, i.e. using one set of deformation parameters for $q < \gamma_0 d$, and another set for $q > \gamma_0 d$*).

In some cases, however, no break in the load-settlement curve has been observed for $q = \gamma_0 d$ (Schultze, 1963 – in all seven structures described; Sherif, 1973, p. 134 – in 48 of 53 structures described). This fact can probably be explained either by overconsolidation of the subsoil during which the overconsolidation load is substantially greater than $\gamma_0 d$ (and/or the behaviour of the subsoil is approximately elastic**), or

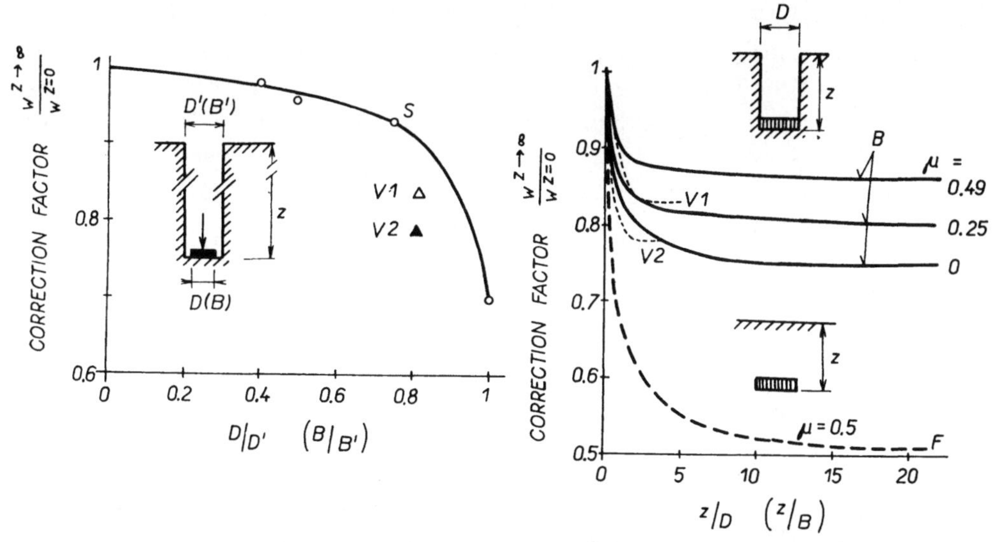

a) b)

Fig. 32. Effect of depth of foundation on settlement ($w^{z=0}$ for surface load. $z=0$; $w^{z\to\infty}$ for deep loading, $z>0$). *a* — effect of relative dimensions of the loaded area: *S* — model tests on araldite, rigid circular plate (Scherman, 1969); V_1, V_2 — strip load on an elastic halfspace, solution by the finite element method for an unlined (V_1) and a lined (V_2) excavation (Vinckurov and Mikulich, 1975). *b* — relation between correction factor, depth of foundation and Poisson's ratio μ: *B* — circular flexible plate in an unlined shaft (Burland, 1969c); V_1, V_2 — strip in an unlined (V_1) and a lined (V_2) excavation not adjoining the excavation edges (cf. a) — Vinokurov and Mikulich (1975); *F* — a uniformly loaded circular area inside a halfspace (Fox, 1948b). An elastic halfspace is considered in all cases. Solutions *B*, V_1 and V_2 were obtained by the finite element method.

---

*) It is difficult to agree with Giroud's (1972, p. 11) procedure. According to him, the settlement would be negligibly small for $q \leqq \gamma_0 d$, and this conclusion is generally unacceptable.

**) Thus, for example, Egorov and Popova (1971) recommend carrying out the calculations for full loading by the structure.

by a disturbance in the subsoil during construction (Wakeling, 1974). After being disturbed, the subsoil "forgets" its previous loading and behaves like a normally consolidated one. Another factor that comes into play is the effect of scale – the break in the load-settlement curve for $q = \gamma_0 d$ is not as distinct for large foundations as is the break in the compressibility curve of small samples (Leonards, 1974).

(c) The effect of the depth of foundation on settlement is usually introduced by means of a correction factor (<1) by which one multiplies the settlement calculated under the assumption that the load acts on the surface. Fig. 32 shows several possibilities of choosing the correction factor. The left-hand side (a) of the figure shows the effect of the relative dimensions of the loading plate according to Scherman's (1969) model tests. These are in good agreement with the calculations of Vinokurov and Mikulich (1975) carried out for a strip. The right-hand side (b) shows the frequently used correction of Fox (1948b)*), Burland's (1969c) curves are, however, more realistic. Although a lining reduces the correction factor (from $V1$ to $V2$), a real subsoil may not be able to take up the tensions which exist in an elastic halfspace. A practical conclusion which can be drawn on the basis of the above is that the minimum correction factor ranges between 0.7 to 0.85 (for $z/D \to \infty$ or $z/B \to \infty$) and in the first approximation is independent of the shape of the loaded area.

3.2 Elastic methods

3.2.1 Homogeneous isotropic halfspace

Elastic methods are most refined for an elastic homogeneous isotropic halfspace; a review of the pertinent solutions is presented, for example, by Florin (1959), Scott (1963), Giroud (1972, 1973) and Poulos and Davis (1974). The publications of these authors contain more complicated solutions than those discussed below, for example, solutions for a subsurface load of a halfspace, various forms of loading areas, etc.

The basis for settlement calculations are the equations of Boussinesq (1885) and Cerruti (1882) defining the vertical displacement w of the surface of a halfspace ($z = 0$) loaded by a vertical (P) and a horizontal (H) load (Fig. 20). The equations are of the form

$$w = \frac{1 - \mu^2}{E} \frac{P}{\pi R} \tag{87}$$

$$w = \frac{(1 + \mu)(1 - 2\mu)}{E} \frac{Hx}{2\pi R^2} \tag{88}$$

*) As implied by a comparison of Kelvin's and Boussinesq's formulae defining the settlement under a concentrated load, the correction factor is $1/2$ ($\mu = 0.5$) to $\sim 1/3$ ($\mu = 0$) for $z/D \to \infty$.

By integrating these equations one can obtain the settlement of loading areas of various shapes (e.g. Steinbrenner, 1934; Schleicher, 1926). Comprehensive calculations of the magnitude of the settlement w at the centre, in the corners and at the centres of the sides of a rectangular area linearly loaded on the surface (Figs. 21 and 22) were presented by Giroud (1968, 1969 – for a vertical load only; 1973). His publication contains the values of all twenty-three coefficients needed in the calculation.

The settlement of a rectangle (circle) subjected to a surface vertical uniform load q is defined by

$$w = \frac{1 - \mu^2}{E} qBI_w \tag{89}$$

where B is the width (diameter) of the rectangular (circular) foundation, I_w – the influence coefficient which depends on the position of the point whose settlement is calculated, on the form of the foundation and on Poisson's ratio (see, e.g. Florin,

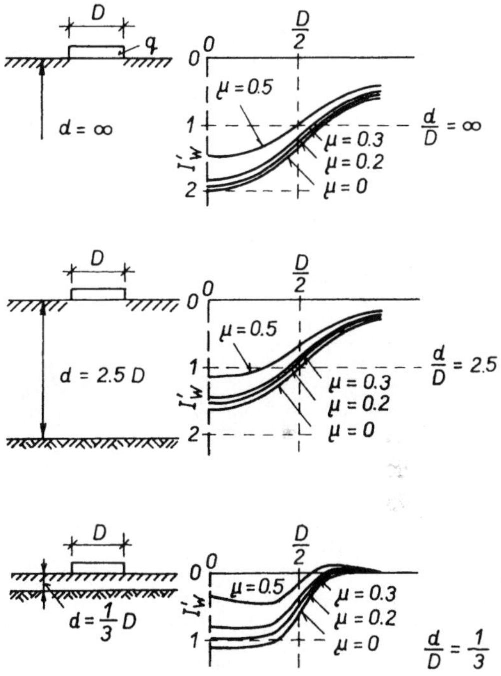

Fig. 33. Influence coefficient for a uniformly loaded circular area (Terzaghi, 1943).

1961, p. 23 – the effect of Poisson's ratio on I_w ceases for the settlement of the surface of a halfspace and for a depth of elastic subsoil $d \to \infty$). If the thickness of the foundation soil is bounded by an incompressible base, the effect of the base can also be included in I_w – cf. Fig. 33 (Terzaghi, 1943). For the centre of a uniformly loaded flexible circle, $I_w = 1$, for a rigid circular foundation, $I_w = \pi/4$. Fig. 33 shows an example of the settlement variations obtained by means of the elasticity theory for a uniformly loaded circular area resting on subsoil of both finite and infinite thickness. The settlement is given by the equation

$$w = q \frac{D}{2E} I'_w \tag{90}$$

(D is the diameter of the circular loading area). Fig. 33 shows how far the settlement of the surface reaches from the loaded area, as well as the possibility of the surface heaving by the effect of load acting on a relatively soft subsoil on an incompressible base.

The settlement of perfectly stiff foundations represents another type of solution with boundary conditions specified in terms of the displacements. The settlement (vertical displacement) of a stiff circular plate was already solved by Boussinesq (1885) – see also Fröhlich (1934, p. 182).

Formulae (87), (88) and those derived from them, for example, eqn. (89), define the settlement as it depends on Young's modulus E and Poisson's ratio μ. In applications to subsoils which do not behave elastically, the term Young's modulus is replaced by the term "the deformation modulus". Formulae (87) and (88) assume a linear stress-strain relation, and E = const., μ = const. within the whole range of the halfspace. In the region of stresses usually encountered in settlement calculations, the value of μ does not differ too much for different soils, the value $\mu = 0.3$ being customarily taken as typical.

The magnitude of the settlement of an elastic linear halfspace subjected to a simple surface load can be obtained by means of the integration

$$w = \int_0^z \varepsilon_z \, dz \tag{91}$$

where $z \to \infty$ for the settlement of the surface of a halfspace not bounded from below. Eqn. (91) in combination with the third of eqns. (13) is used by Poulos and Davis (1968) to obtain directly the final settlement by replacing integration by summation; a similar procedure was already used by Vargas (1961). Fig. 34 shows the contribution of the lower layers to the magnitude of the settlement in the case of an elastic halfspace, uniformly loaded on a circular area.

Eqn. (91) can be applied generally, i.e. also to an inelastic medium. The variations of ε_z with depth of an elastic halfspace uniformly loaded on a circular area on the surface can be determined by using the formulae of Ahlvin and Ulery (1962). It is found that ε_z increases with depth (for $\mu = 0.5$, $\varepsilon_z = 0$ for $z = 0$), reaches its maximum for $z/D = 0.3$ to 0.35 (Schmertmann, 1970; Lambe and Whitman, 1969, p. 213) and then decreases. If a foundation soil deforms only vertically, ε_z decreases monotonically with depth, as implied by eqns. (42) and (50) (Fig. 60).

A variant of the classical solutions of the elastic methods is the method of equivalent layer introduced by Tsytovich (1963).

The settlement of an elastic halfspace subjected to a uniform rectangular subsurface load is frequently obtained with the help of a diagram published by Fox (1948b), as already mentioned in connection with Fig. 32. When the load is applied at a great depth, the settlement can decrease all the way to half the settlement under surface loading; when $\mu < 0.5$, the decrease can even be greater (for a circular area and $\mu = 0$, the correction factor is equal to $8/3$ – Nishida, 1966). Fox's diagram shows

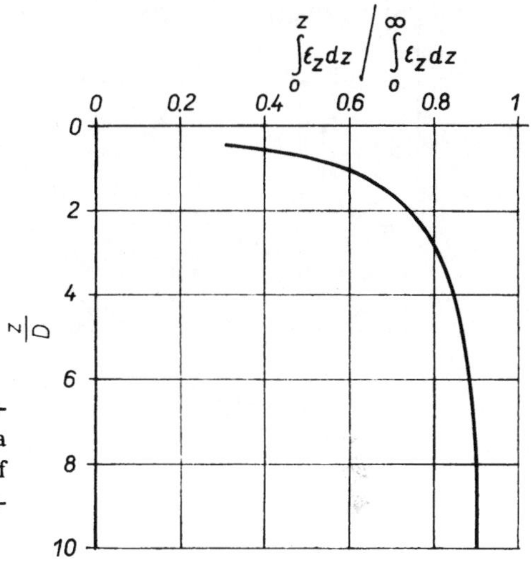

Fig. 34. Effect of neglecting subsoil compression below certain level on the settlement of a uniformly loaded circular area on the surface of an isotropic elastic halfspace (Lambe and Whitman, 1969).

the average settlement which does not differ too much from the settlement of a rigid foundation. The settlement of both flexible and rigid rectangular subsurface foundations can be determined using the diagrams of Groth and Chapman and Butterfield and Banerjee (cf. Poulos and Davis, 1974, p. 92 and 180). In these cases, it is, however, more realistic to apply the correction factor shown in Fig. 32.

As early as 1897, Föppl observed discrepancies between experiments and the theory of linear elasticity so far as the settlement of the surface outside of a loaded area is concerned. Actual vertical displacements attenuate far more rapidly with the distance from the loaded area. Föppl ascribed the disagreement to the invalidity of Hooke's law in soils (Fröhlich, 1934, p. 10). Investigations have, therefore, been made of the possibilities of calculating the deformations of elastic halfspaces having different mechanical properties.

3.2.2 Cross-anisotropic halfspace

For cross-anisotropic halfspaces, Barden (1963) showed (with a certain simplification – cf. Section 2.4) that a simple relation can be found between the settlement of a loaded area on an isotropic halfspace (w_{isot}) and that on a cross-anisotropic halfspace (w_{anisot}), and expressed the anisotropy by means of a dimensionless factor (Fig. 35). It is clearly seen in this figure that for $n > 1$ (for example, overconsolidated clays), the settlement is smaller than in an isotropic halfspace; hence the distribution ability of the subsoil has increased. An example of the general solution of Koning

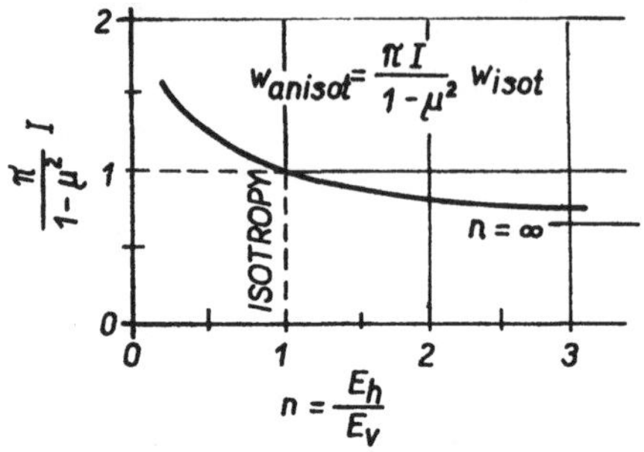

Fig. 35. Relation between the settlements of an isotropic and a cross-anisotropic halfspace (Barden, 1963).

(1957) is shown in Fig. 36. As the figure indicates, theoretically the settlement of an anisotropic halfspace can even be larger than that of an isotropic halfspace (see also Gibson, 1974). In practical calculations of a linearly loaded strip and circle, one can also use the tables of Gerrard and Harrison (1970a,b).

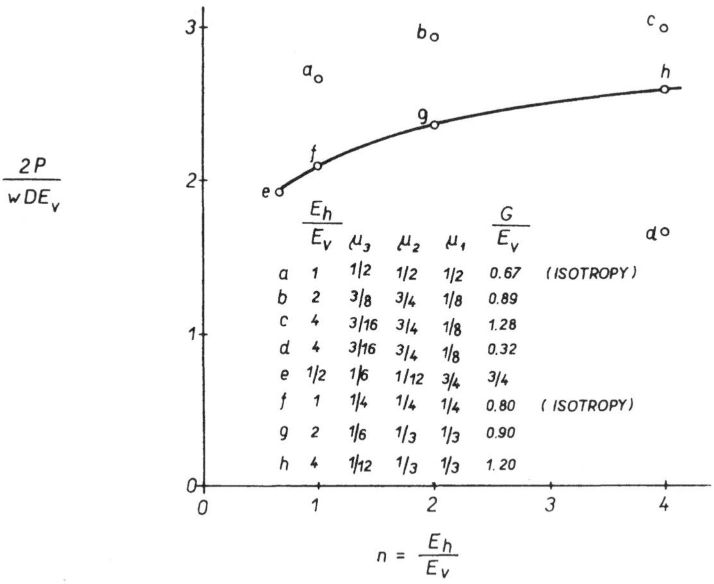

Fig. 36. Settlement of a rigid circular plate on a cross-anisotropic halfspace of different properties — Koning (1957). (P — total load applied to the plate).

3.2.3 *Non-homogeneous halfspace*

According to Hruban (1945), an approximate solution for a continuously non-homogeneous halfspace defined by eqn. (63) can be obtained under certain simplifying conditions (Feda, 1960, 1963). Since the non-homogeneous halfspace is linear, the principle of superposition applies. The settlement can be determined by superposing Boussinesq's solution for a homogeneous halfspace upon Borowicka's solution for a completely heterogeneous halfspace (defined by eqn. (61) – cf. Belloni and Jamiolkowski, 1973). A solution for the case of non-homogeneity defined by eqn. (61) is comparatively simple. If $m = 0.5$, the settlement of the centre of a uniformly loaded (load q) circular area of diameter D is given by (for z_0 see (61))

$$w = \frac{28q\sqrt{(D.z_0)}}{15E_0\sqrt{2}} \tag{92}$$

It can be shown (Hruban 1945) that the increase of the deformation modulus with depth accounts for the fact that 9/10 of this settlement is a result of subsoil compression to the depth $2D$.

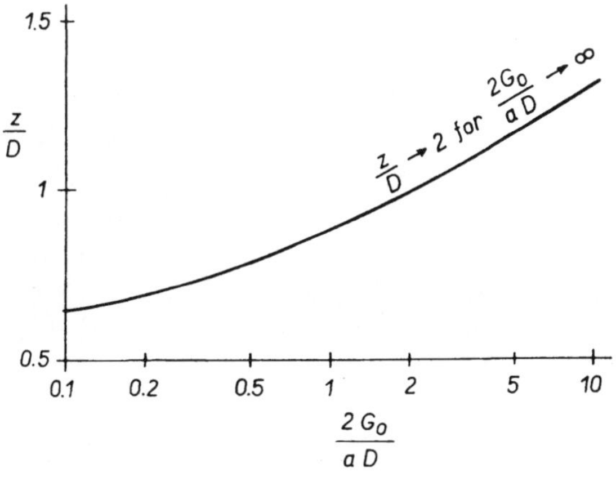

Fig. 37. Depth of concentration of strains (active zone) in an incompressible, non-homogeneous (cf. eqn. (69)) halfspace uniformly loaded on a circular area (after Gibson and Som — cf. Garga and Quin, 1974).

From the requirement that the settlement (to the depth $z = 2D$) of a homogeneous halfspace should be the same as that of a non-homogeneous halfspace defined by eqn. (69), one can calculate the depth (of the so-called active zone) to which the deformation propagates in the latter case (Fig. 37; see also Feda et al., 1977). Due to the effect of subsoil non-homogeneity, the deformation clearly concentrates immediately beneath the loading area. As Burland and Wroth (1974) point out, 80% of the settlement of a circular foundation is a result of the deformation of the subsoil from the surface to a depth of only $0.6D$. A similar conclusion is reached by Awojobi (1975) for the case of subsoil whose non-homogeneity is defined by eqn. (70). As will be shown in Section 3.10, this is in agreement with numerous observations.

An interesting result concerning a halfspace which is non-homogeneous in accordance with eqn. (61) for $m = 1$, or with eqn. (73), i.e. $E = 0$ on the surface which is also incompressible ($\mu = 0.5$ by eqn. (66)), was obtained by Gibson (1967) for both an axially symmetric problem and the case of plane deformation under a uniform load q. In this case the settlement is uniform inside the loaded area and zero outside it, and it holds that

$$w = \frac{q}{2\beta_0} \tag{93}$$

if $E = 3\beta_0 z$. Eqn. (93) can be re-written in the form

$$q = wk_s \tag{94}$$

which is identical with that for a Winkler halfspace (Winkler, 1867). The coefficient of the vertical reaction of the subsoil, k_s, is constant and defined as

$$k_s = 2\beta_0 \tag{95}$$

It can be concluded on the basis of the above that solutions founded on a constant coefficient of the vertical reaction of the subsoil, k_s, independent of the magnitude of the loading area, are exact for incompressible, continuously non-homogeneous and deep (i.e. the thickness of the subsoil is greater than the width of the foundation – Gibson et al., 1971) subsoils to which applies eqn. (61) and $m = 1$. The condition that the subsoil must be deep enough relates to a rough contact with the incompressible layer. Numerical calculations made for the case of a smooth contact have shown that the theoretical prediction is sufficiently accurate even for a layer whose thickness is equal to 4% of the foundation width (Simons and Rodrigues, 1975). If $\mu \neq 1/2$ and/or $G_0 \neq 0$ in eqn. (69), eqn. (94) no longer applies (Brown and Gibson, 1972). Calculations of the settlement of a loaded rectangle and circle on a non-homogeneous subsoil described by eqn. (69) are facilitated by using diagrams prepared by Brown and Gibson (1972, 1973). The influence coefficients for the settlement of a rigid smooth circular plate were calculated by Carrier and Christian (1973). All the calculations generally show that the settlement substantially decreases due to the effect of subsoil non-homogeneity.

In his calculations of the settlement of a continuously non-homogeneous layer of subsoil (the non-homogeneity is defined by eqn. (63) for $m = 1$, or by eqn. (69)) on an incompressible base, Butler (1974) used the method developed by Steinbrenner and published diagrams (Figs. 38 to 40) which are convenient for engineering calculations. They were completed by Meigh (1976). The thickness of the compressible layer can also be interpreted as the thickness of the so-called active zone.

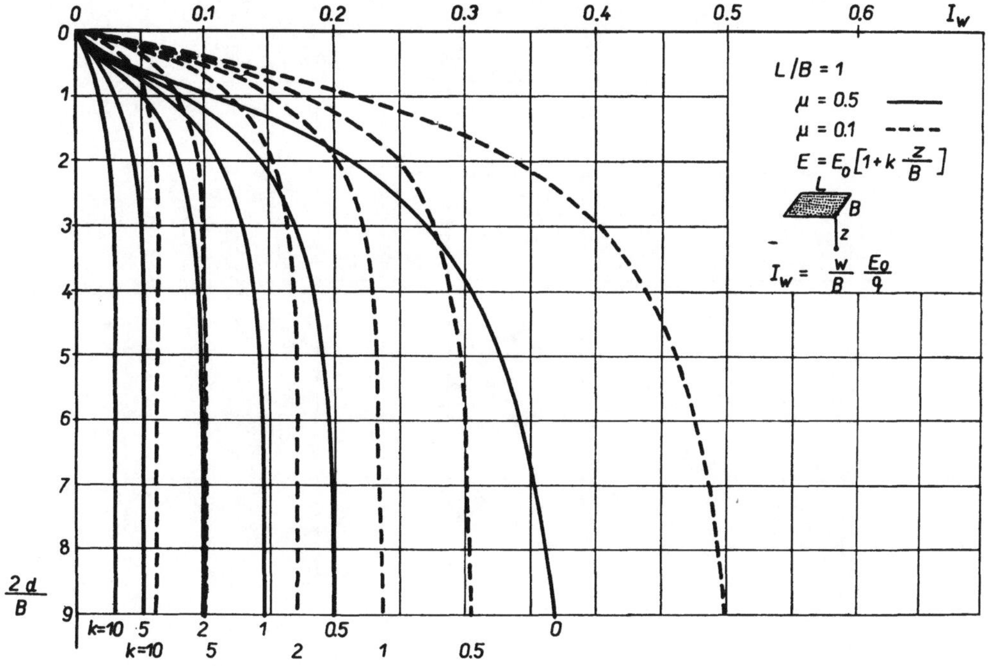

Fig. 38. Diagram for calculating the settlement of a square area uniformly loaded by q on a non-homogeneous subsoil of thickness d (for $\mu = 0.1$ and 0.5) (Butler, 1974).

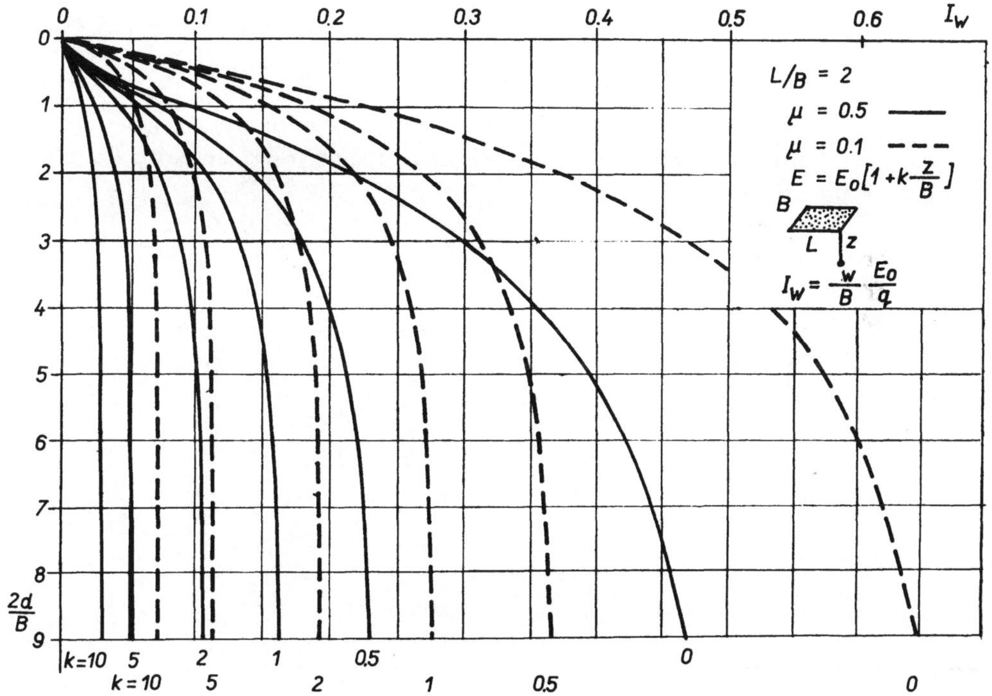

Fig. 39. Diagram for calculating the settlement of a rectangular ($(L/B) = 2$) area uniformly loaded by q on a non-homogeneous subsoil of thickness d (for $\mu = 0.1$ and 0.5) (Butler, 1974).

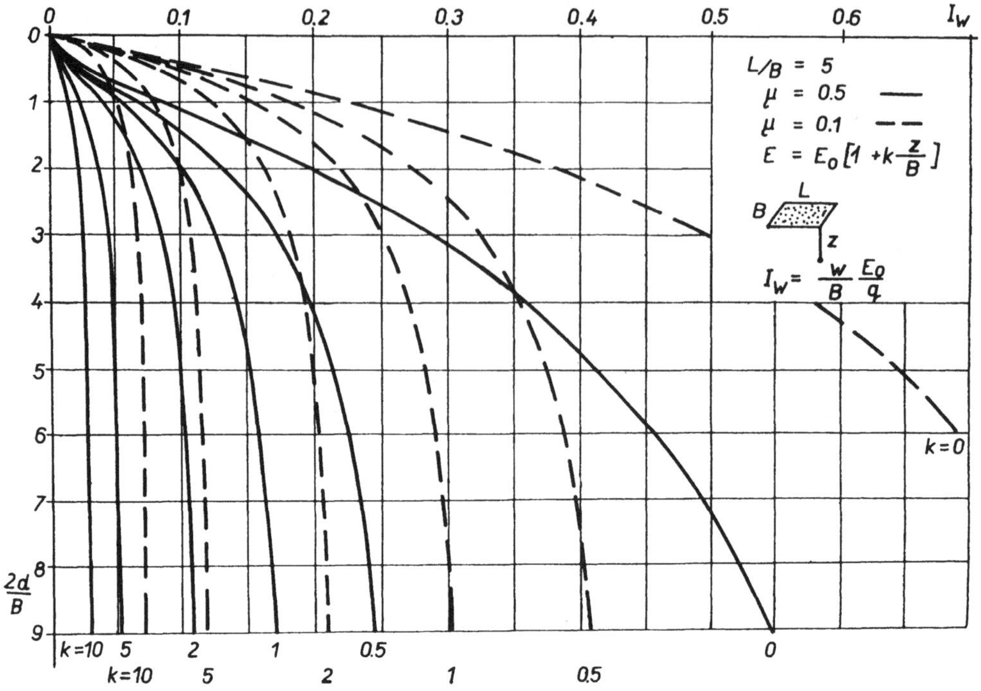

Fig. 40. Diagram for calculating the settlement of a rectangular ($L/B = 5$) area uniformly loaded by q on a non-homogeneous subsoil of thickness d (for $\mu = 0.1$ and 0.5) (Butler, 1974).

If the subsoil consists of layers with different types of non-homogeneity, the variations of the deformation modulus with depth can be superposed upon one another as shown in Fig. 41 (Butler, 1974).

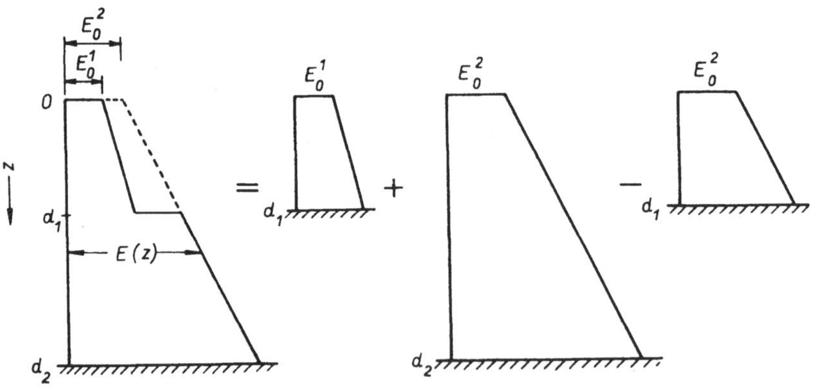

Fig. 41. Superposition of deformation moduli of non-homogeneous subsoils (Butler, 1974).

An analysis of the deformation of a non-homogeneous halfspace makes it possible to formulate a conclusion which is generally of essential significance for any settlement calculation. As put forward in Section 2.5.1, the states of stress of an incompressible homogeneous and of an incompressible non-homogeneous (according to eqn. (73)) halfspace are identical. However, as a comparison of Fig. 33 and eqn. (94) shows, the deformations of the two halfspaces considerably differ from one another. Hence, if the physical model of the subsoil can often radically be simplified in stress calculations, this procedure cannot be allowed in deformation calculations. As will be shown later, a similar conclusion also applies to a non-linear halfspace.

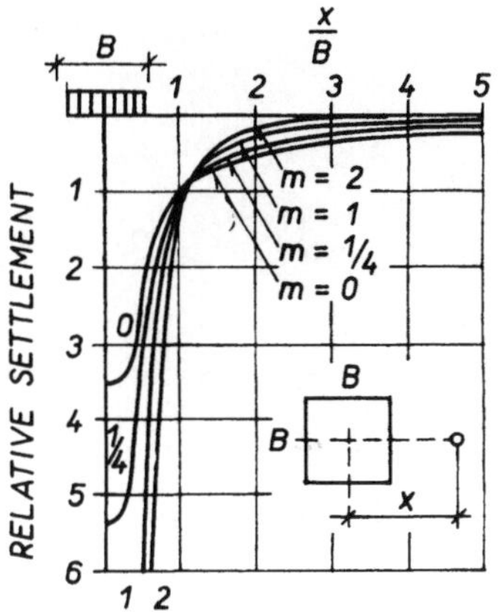

Fig. 42. Settlement of a square loading area for different rates of increase of Young's modulus of soil with depth (Klein, 1956).

It is found that in the case of a continuously non-homogeneous halfspace whose deformation modulus is supposed to increase with depth, the vertical displacements of the surface attenuate with the distance from the loaded area the faster, the larger is the value of m (a similar effect is also noted for decreasing G_0/m ratio when the non-homogeneity is defined by eqn. (69) – Brown and Gibson, 1972). This is clearly seen in Fig. 42 relating to a square and a subsoil described by eqn. (61). If $m \geqq 1$, then the settlement of the centre, w, tends to infinity (in plane deformation, $w \to \infty$ only for $m = 0$). It should be borne in mind that eqn. (61) is not realistic in practical applications because E is non-zero even on the surface of a real foundation soil. This is the reason why more important by far is the non-homogeneity described by eqn. (63), or by eqn. (69) in the simplest case of $m = 1$. However, a solution can only be obtained by means of numerical methods (Biarez and Boulon, 1971; Brown and Gibson, 1972, 1973).

As experimental results relating to sand and eqn. (63) suggest, the range of m is $0.25 \leqq m \leqq 0.75$ (Feda, 1960, 1963); field observations most frequently yield the value of $m = 0.5$ to 0.8 (see, e.g., Čermák and Klapetek, 1959; Feda, 1960, p. 119; Mencl, 1966, p. 190). For m in eqn. (61), Lobanov (1965) proposes the values 0 to 0.4 (clays) and 0.4 to 0.8 (sands). In keeping with eqn. (65), $\mu = 0.4$ to 0.5 (clay) and $\mu = 0.35$ to 0.4 (sand) would correspond to these values.

3.2.4 *Layered halfspace and an incompressible base*

A discontinuously non-homogeneous (layered) halfspace can be solved, for example, by the method of Steinbrenner (1934) which consists of integrating eqn. (91) within the range of each layer (on the assumption that the stress propagates just as in a homogeneous halfspace – cf. Section 2.5.2) and adding up these partial deformations to obtain the resulting settlement. Other methods have been proposed by Poulos and Davis (1974, p. 162 et seq.). If the contact between the layers is smooth, the settlement increases up to twofold but this effect can only be observed for a perfectly smooth contact (Novotný, 1975).

Numerous solutions are available for the case when the foundation soil of fairly small thickness is bounded by an incompressible base. One of them is shown in Fig. 33.

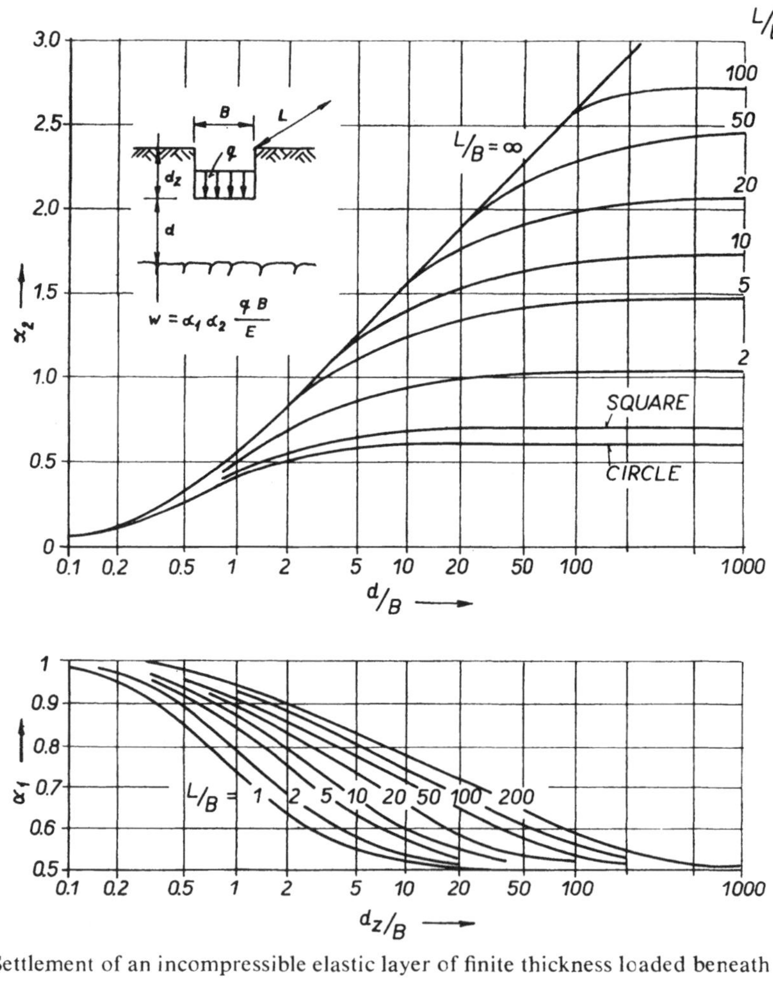

Fig. 43. Settlement of an incompressible elastic layer of finite thickness loaded beneath its surface (Janbu et al., 1956).

Solutions in which a rough contact with the incompressible base is considered (Burmister, 1956; Milovic and Tournier, 1971; Poulos, 1968) are more realistic than those assuming a smooth contact (Sovinc, 1961, 1969). The difference between these two solutions is the greatest for $\mu = 0.5$ (Ueshita and Meyerhof, 1968). Hobbs (1974) finds considerable differences between the solutions of various authors, which are probably caused by the necessity of using numerical methods. The solution of Gorbunov-Posadov (1953, p. 495; see also Tsytovich, 1963, p. 468) shows the settlement of the centre of a uniformly loaded area to be independent of the shape of the area provided that the thickness of the compressible layer is less than a half of the width of the loaded area (a rough contact with the incompressible base). A solution for a continuously non-homogeneous compressible layer was published by Gibson et al. (1971). For the case of a smooth contact, Gibson (1974) draws a surprising conclusion that the settlement of a strip is independent of the thickness of the compressible layer. This was essentially proved by the solution obtained by means of the finite element method (Simons and Rodrigues, 1975).

Practical calculations can make good use of a diagram published by Janbu et al. (1956) and reproduced in Fig. 43. This diagram enables the calculations to take account also of the effect of the depth of foundation. In practical calculations the thickness of the deformable layer can be identified with the depth of the so-called active zone (cf. Section 3.5). Just as in the solution of Fox (1948b) on which the diagram is based, the calculated settlement is the average settlement of a foundation and applies to an incompressible subsoil ($\mu = 0.5$).

3.2.5 Non-linear halfspace

The effect of a concentrated load can no longer be simply integrated when deriving the settlement of a loaded area for a non-linear halfspace to which the principle of superposition does not apply. An approximate relation obtained for this case on the basis of Hrubans results is given by Feda (1960, 1963). The solution is insignificant also for the reason that for a real foundation soil, the exponent k in eqn. (78) cannot be taken for a constant independent of depth. Laboratory tests yielded the values $k = 1.5$ to 2 for sand (Feda, 1960, 1963), and in the case of eqn. (82) the values $k = 0.5$ to 0.6 for sand in a triaxial apparatus (Biarez, 1962, p. 63).

Huang (1968) calculated the settlement of a circular, uniformly loaded plate resting on a subsoil whose non-linearity is described by eqn. (81). Although the differences in the stress σ_z amount to 30% max., the settlement in the non-linear case can be as little as 1/4 of that in the linear case ($E = E_0 =$ const.). This clearly demonstrates the significance of the non-linear stress-strain relation for the calculation of settlement, as compared to the calculation of stresses in which the non-linearity can often be neglected. Similar results were obtained by Biarez and Boulon (1971) also using numerical methods. In a non-linear halfspace one finds a similar concentration of deformations immediately beneath the loading area as in a non-homogeneous halfspace (Fig. 37). 94% of the settlement of a circular plate of diameter D concentrates to a depth of $1.2D$ (Hruban, 1958) or possibly, of $1.5D$ (Shirokov et al., 1971). In a linear halfspace, only about 2/3 of the settlement takes place to this depth. Vargas (1961) calculated the settlement of seven tall buildings on sand for a halfspace which was non-homogeneous as well as non-linear: the deformation modulus was a function of the sum of the horizontal geostatic stress and the stress resulting from additional loading of the subsoil.

In consequence of the principle of superposition, the elasticity theory does not explicitly consider the effect of the dead weight of the medium (which is introduced indirectly as non-homogeneity, anisotropy or non-linearity). On the basis of an analysis of the plane problem of elasticity theory, Gorbunov-Posadov (1972) concludes that the solution is not unique because the displacements are too large. The condition of minimum potential energy which must be satisfied by correct solutions, also contains the unit weight which has a dominant effect under light loads; under heavy loads the conditions approach those of a weightless medium. According to Gorbunov-Posadov, this different effect of the unit weight leads to a non-linear load-settlement relation even when the linear theory is applied. This is a remarkable conclusion which can be supplemented by a statement of an opposite kind, i.e. that the nonlinear theory is also apt to yield a linear load-settlement relation (cf. Section 3.10, Fig. 49 – Kryzhanovskiy et al., 1975).

3.3 The stress-path method

The idea of this method, which is also general in other engineering disciplines, is logical: to derive the mechanical parameters and deformations of undisturbed samples of subsoils by a laboratory test during which the stress variations approach most closely the actual variations in a standard element of the subsoil in the field. Terzaghi (1925) regarded the oedometer test in which lateral deformation of the sample is practically completely prevented from taking place as such a test for settlement calculations (first oedometric tests were performed by Frontard as early as 1914).

The method was worked out, justified and widely applied mainly by Lambe (1963, 1964, 1967), and independently of him, proposed by Davis and Poulos (1963). The stress path is simulated in a triaxial apparatus. The stress path is a line connecting the points which express the successive changes of stress during additional loading or unloading in the (p, q) plane – the total stress, or in the (p', q) plane – the effective stress. The pertinent relations are:

$$p = \frac{\sigma_v + \sigma_h}{2}$$

$$p' = \frac{\sigma_v' + \sigma_h'}{2}$$

$$q = \frac{\sigma_v - \sigma_h}{2} \tag{96}$$

where σ_v, σ_h, σ_v', σ_h' are the vertical and the horizontal normal total and effective stresses (using notation of previous section, $\sigma_z = \sigma_v$ and $\sigma_x = \sigma_y = \sigma_h$). In the discussion that follows, a single quantity will be used, namely p, which becomes $p = p'$ when an effective stress path is considered.

In Mohr's plane (τ, σ_n) each state of stress at a point of a halfspace is characterized by a single circle whose centre and radius are determined by quantities p and q. Since in the oedometric test $\sigma_h/\sigma_v = K_0 = \text{const.}$, one obtains

$$\frac{p}{q} = \frac{1 + K_0}{1 - K_0} = \text{const.} \tag{97}$$

and the stress path is a straight line passing through the origin of the (p, q) plane. To this path corresponds a uniaxial deformation of the sample; the transverse deformation is zero. The typical stress paths in the (p, q) plane for a constant stress increment $\Delta\sigma_z$ (hence the angle $EFB = 45°$) and a schematic representation of the types of sample distortions are shown in Fig. 44. In Fig. 44a, the path EF denotes isotropic (hydrostatic) compression, the path AC — the uniaxial (oedometric) deformation; since AB produces an extension of the sample, the axial deformation is larger than for AC; AD — the axial deformation is smaller than for AC. In Fig. 44b the initial state of the sample is isotropic consolidation; AB is the stress path under

a)

b)

c)

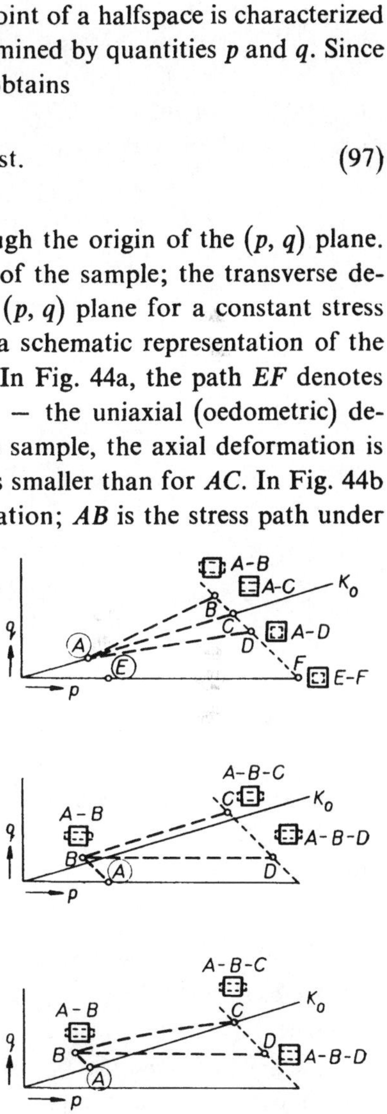

Fig. 44. Typical stress paths for triaxial tests (Lambe, 1964) a — distortions from drained compression; b, c — distortions from undrained compression followed by drained compression; letters in circles — initial state of stress.

shear loading and constant volume of the sample; during consolidation along *BD*, both vertical and transverse compressions of the sample take place; during consolidation along *BC* the transverse compression is comparatively slight. Fig. 44c – the initial state of the sample is a result of anisotropic (oedometric) consolidation; *AB* – shear loading under constant volume, the transverse deformation of the sample is smaller than that for the path *AB* in Fig. 44b; during consolidation along *BD* the transverse compression of the sample is such that *ABD* gives zero transverse deformation; *BC* – slight transverse compression which is incapable of eliminating the transverse extension from the phase *AB*.

The stress paths for a typical element of a subsoil subjected to a load are shown in Fig. 45. The in situ geostatic state of stress is represented by the point *I*, the final state, after consolidation has been terminated, by the point *F* (both in effective stresses). The shape of the stress path between the points *I* and *F* depends on the nature of the subsoil and on the character of the load.

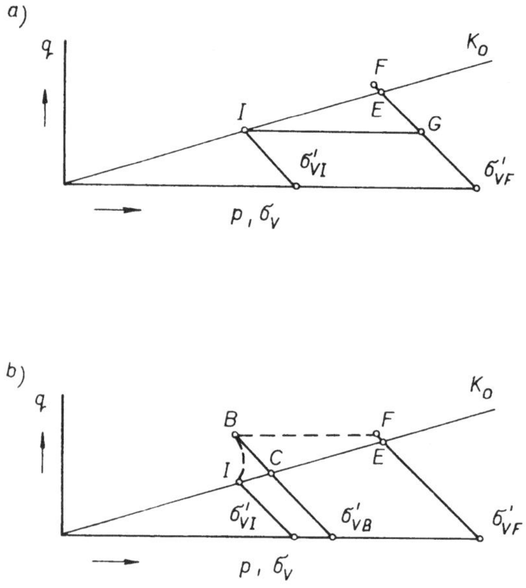

Fig. 45. Typical stress paths for settlement calculations (Lambe, 1964).

The procedure used in settlement calculations by means of the stress path method can be summarized into the following points (after Lambe and Whitman, 1969, pp. 216 and 489):

(1) Selecting one or more representative points in the layers of the subsoil, which contribute to the settlement.

(2) Determining the in situ stress (including the neutral stress) at these points (the point *I* in Fig. 45).

(3) Calculating the changes in the total stress produced by the effect of load at each chosen point; in the case of a two-phase half-space, separately for the initial

phase – without draining, at a constant volume of the subsoil – (the path *IB* in Fig. 45b), and separately for the subsequent consolidation (the path *BF* in Fig. 45b).

(4) Performing laboratory tests which follow the stress path from the initial to the final state of stress and during which the deformations of the sample are measured. In two-phase halfspaces, the sample is first loaded without the possibility of draining and then left to consolidate to the calculated state of stress.

(5) The vertical deformations obtained by measurements according to point (4), are used in settlement calculations; in the case of a two-phase halfspace, separately for the immediate and separately for the consolidation settlement.

(6) In the case of a two-phase halfspace, the time necessary for consolidation is measured and used in calculation of the time variation of the settlement.

A typical result of calculations is shown in Fig. 46. The magnitude of the vertical strain ε_v $(=\varepsilon_z)$ was determined along a vertical line beneath the centre of the tank

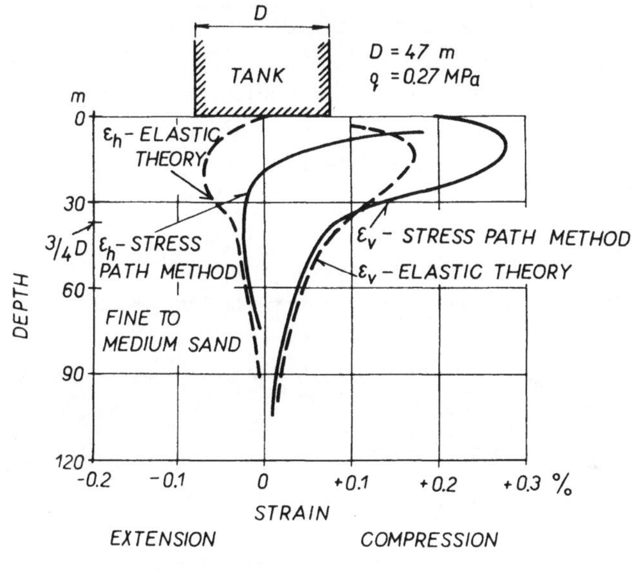

Fig. 46. Vertical (ε_v) and horizontal (ε_h) strains in a subsoil loaded by a circular tank – solution by the application of the elasticity theory and the stress path method (Lambe, 1967; Lambe and Whitman, 1969).

(of about 47 m dia., subjected to a load of about 0.27 MPa) at several points which were used to construct the curve $\varepsilon_z = \varepsilon_z(z)$. The magnitude of the settlement w can be obtained by numerical integration of this curve.

Notes on some points of the calculation:

On point (1): if the pattern bounded by the curve $\varepsilon_v = \varepsilon_v(z)$ and by the vertical centreline of the foundation (Fig. 46) were replaced by a rectangle, the point of intersection of the substitute and the actual patterns would determine "the average point". If ε_z at this point is known, the settlement is calculated by merely multiplying ε_z by the thickness of the compressible layer. The choice of the average point is difficult (for a circular loading area it lies at a depth approximately equal to 3/4 of the diameter – Lambe and Whitman, 1969, p. 217); hence many points are frequently needed to obtain the curve $\varepsilon_z = \varepsilon_z(z)$.

On point (3): The change in stress is calculated by means of the theory of linear elasticity. To determine a point of stress in the (p, q) plane, one must know σ_z $(=\sigma_v)$ as well as σ_x $(=\sigma_h)$; usually, one is dealing with an axi-symmetrical or a plane problem. In the calculation one must use Poisson's ratio (in the case of plane deformation problems, only when the boundary conditions are specified in terms of the displacements). In two-phase subsoils, $\mu = 0.5$ (incompressible subsoil) is employed for the first stage of loading (immediate settlement), and $\mu' \doteq 0.3$ for the second stage (consolidation settlement).

On point (4): The stress path is determined from the results of triaxial and oedometric tests (one can also measure the stress paths when the sample is under plane deformation, and exceptionally, even general stress paths when $\sigma_x \neq \sigma_y$). Hence the actual stress path is only followed with a certain degree of approximation but the possibility of using standard apparatuses is an advantage. According to Fig. 45b, a typical stress path consists of the section *IB* – for which the vertical strain of the sample, ε_{v0} is taken directly from an undrained triaxial test, and of the section *BF* – consolidation stress path which usually approaches the radial stress path. For the radial stress path the volume strain ε_{vol} of the sample can be ascertained from the compression of the sample in the oedometer, i.e. for K_0 – consolidation. If ε_{vol} is known, the axial compression of the sample in the consolidation phase, ε_{vc}, is obtained from the relation (Lambe, 1964; Seed, 1965)

$$\varepsilon_{vc} = \frac{1 + K_0' - 2K'K_0'}{(1 - K_0')(1 + 2K')} \varepsilon_{vol} \tag{98}$$

where K_0' and K' is, respectively, the initial and the final ratio of the effective stresses σ_h' and σ_v'. Eqn. (98) was derived on the basis of the elasticity theory of isotropic media (Šuklje, 1969, pp. 379 and 380). Simons and Som (1969) who propose using the theory of cross-anisotropic media instead, show with the help of experiments that it leads to a better agreement.

A version of the stress path method combines this method with the theory of isotropic linear elasticity. The calculation of the instantaneous and consolidation settlements is carried out according to the elasticity theory while the elastic (deformation) modulus is obtained from the stress path in the (q, ε_v) plane. For the final stage (after the termination of primary consolidation) one uses the parameters E' and μ' $(\doteq 0.3)$, for the initial (instantaneous) settlement, the parameters E and $\mu = 0.5$. The value of E is sometimes calculated from the relation

$$E = \frac{3}{2(1 + \mu')} E' \tag{99}$$

(derived for linear and isotropic media), or is determined experimentally from the stress-strain diagram of a standard compression triaxial test when $\sigma_x = \sigma_y = \sigma_r =$ $=$ const. (cf. the last of eqns. 13) as

$$E = \frac{\Delta\sigma_v}{\Delta\varepsilon_v} \tag{100}$$

In this procedure, Hooke's law is replaced by a real quasi-linear relation, and the deformation modulus is the secant modulus (Lambe and Whitman, 1969, pp. 435 and 488, etc.).

The stress path method is a logical method although it is not rigorously correct from the mathematical point of view, for it is a combination of the linear elasticity theory (more complicated theories can also be used) and deformation determined by a direct test of samples with a prescribed stress path. Thus two constitutive relations are employed simultaneously, i.e. Hooke's law for the stress calculation and an experimental relation for the strain calculation. This is, of course, typical of all methods of the second group (cf. Section 3.5 and Section 3.6).

Of all the methods of settlement calculations, the stress path method expresses the mechanical behaviour of subsoils most realistically. This is by far the most important and the most difficult task of any method of settlement calculations. One proceeds from a wholly general assumption that the deformation, neutral stress and strength of subsoils depend on the stress path (Lambe, 1967); the procedure makes it possible to gain a more thorough understanding of the processes which take place in situ when a foundation soil is loaded by a structure.

The disadvantages of the stress path method can be summed up as follows: considerable amount of work and time needed for the experiments (which call for a high quality of undisturbed samples and excellent standard of laboratory tests); uncertainty associated with the choice of "the average point"; and impossibility of exact laboratory reproduction of the actual stress path which is a priori assumed to be independent of the properties of the subsoil.

3.4 The state boundary surface method

Just as the stress path method, the state boundary surface method, described and experimentally verified by Burland (1969a, 1969b, 1971) and Wroth (1977), starts from the state of stress in a linear halfspace (with certain modifications). The deformation is calculated by means of other constitutive relations based on the theory of the state boundary surface of normally consolidated clay elaborated by the Cambridge group of Professor Roscoe and his co-workers (a review of the theory is presented by Roscoe, 1969; the last version of the theory, by Roscoe and Burland, 1968; the elementary fundamentals by Schofield and Wroth, 1968).

Although – like the stress path method – it is not entirely correct mathematically, the state boundary surface method has been found acceptable on the strength of its comparison with experiments. In contrast to the stress path method, deformation calculations made by its means employ a theoretical relation between stress and strain (elasto-plastic, or, to simplify, plastic) whose determination calls for the knowledge of several fundamental mechanical parameters of the subsoil (in the least complicated case, of the compression index C_c from an oedometric test – cf. eqn. (107) – and of the effective peak angle of internal friction, Φ_f').

Theoretically, the method is only justified for normally consolidated (or possibly, slightly overconsolidated) clays when the peak angle of internal friction is identical with the residual angle, and the volume changes are negative (volume reduction) or zero (contractantial behaviour). It was applied to the solution of practical problems of plane deformation (foundation strips, embankments).

The calculation of the state of stress allows for the existence of local plastic regions in the subsoil. The vertical stress, which – according to the findings of Hoëg et al. (1968) and Morgenstern and Phukan (1968) – is virtually not influenced by the plastic regions, is then obtained on the basis of the theory of a linear halfspace; the other two components of the stress tensor are calculated from the condition of strength and of invariability of the direction of the axes of principal stresses (Burland, 1971).

For the stress increments produced by the load increments one then calculates the corresponding strain increments at various points of the subsoil using the constitutive relations derived from the state boundary surface. This surface is based on empirical findings (for example, on the logarithmic law of volume compressibility), on a certain assumption concerning the energy dissipation during the deformation process, on an assumption concerning the existence of a yield surface with isotropic hardening (if the point which represents the state of stress in the stress space lies beneath this surface, only a reversible deformation takes place; if it lies on it, the deformation is irreversible, plastic; if it lies above it, hardening occurs; the directrix of the yield surface is the curve of constant porosity*) and hence it corresponds to the results of undrained tests of a water-saturated sample), and on the validity of the normality law and of some other properties of normally consolidated clays (zero effective cohesion, existence of the critical state line).

---

*) This is the case in Burland's (1971) simplification which ignores the reversible (elastic) strain components. The so-called "elastic wall" is then a plane parallel to the plane of the coordinates (p, q); in a general case, non-linear elasticity is assumed (Roscoe and Burland, 1968), and the "elastic wall" is a curved surface.

Fig. 47 shows a comparison of measured and calculated strains of a model foundation strip ($B = 2.7$ cm) on clay. The calculations were made by means of both the general theory – the state boundary surface method (dash lines) and the oedometric compression method (dot-and- dash lines – cf. Section 3.5); so far as the vertical displacements are concerned, Fig. 47 shows only the results of calculations made for the last loading stage. The comparison shows a very good agreement between the theory and

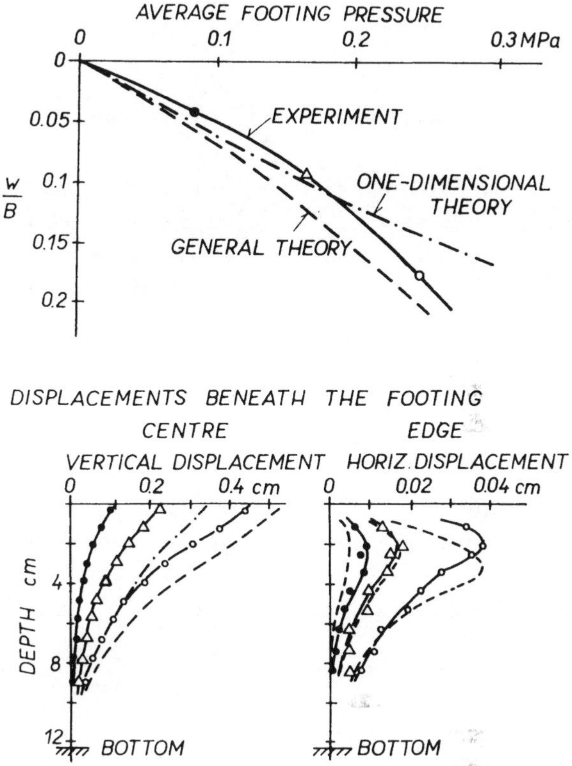

Fig. 47. Vertical and horizontal displacements of a model subsoil (clay) beneath a model foundation strip (Burland, 1969b).

the experiments. A still better agreement was achieved by further refinement of the calculations (the compression index C_c was chosen in a such way as to make the settlement of the foundation during the first loading stage correspond to the calculated one – Burland, 1971). The state boundary surface method can also be used to determine the magnitude of the pore pressure (cf. Section 2.2.1), of hardening by the effect of filtration consolidation, etc. (Burland, 1971).

The advantage of this theory is the fact that all that is needed for its application is the knowledge of several fundamental mechanical parameters of the subsoil, which can easily be measured. A shortcoming is a comparatively narrow applicability of the method. The basic theory of the state boundary surface method was only elaborated in detail for normally consolidated clays. Its extension to other soils meets with considerable difficulties (Roscoe, 1969). It is still somewhat uncertain to what extent one can use the fundamental relations for undisturbed samples (the relevant experiments were almost exclusively carried out with laboratory-prepared re-consolidated samples).

A major contribution of this method is the fact that it is the only one of all methods now in use which applies the constitutive relations that pay heed to the contractantial behaviour of soils (coupling between shear and volume strains – Burland, 1969a). It is in this respect that the method is valuable.

3.5 The method of oedometric compression

This method was evolved and refined by Terzaghi (1925). It is assumed that compression of the subsoil takes place without transverse deformations and that consolidation is uniaxial. Hence the method is used for consolidation settlement calculations. In two-phase halfspaces, the increment of the neutral stress at the instant of a load increase is $\Delta u = \Delta\sigma_z$. The following relation is used in the calculations:

$$w_{oed} = \int_0^z \Delta u \, m_v \, dz \tag{101}$$

(m_v – the modulus of volume compressibility); the integral is usually replaced by summation. Eqn. (101) corresponds to eqn. (91) for ε_z defined by eqn. (42).

The relations usually proposed in connection with this theory are (for $\Delta u = \Delta\sigma'_z$ – an increase of load acting on the subsoil and produced by a structure; the unloading due to excavation is, as a rule, subtracted from the structure load):

$$w_{oed} = \sum_1^i \Delta\sigma'_z \, m_v \, \Delta h_i \tag{102}$$

($\Delta\sigma'_z$ usually corresponds to the theory of linear homogeneous isotropic halfspace; Δh_i – the thickness of the layers into which the subsoil is divided for the purposes of calculation),

$$m_v = -\frac{\Delta e}{1 + e_0} \frac{1}{\Delta\sigma'_z} \tag{103}$$

(e_0, Δe – the initial void ratio and its change produced by the increment $\Delta\sigma'_z$),

$$w_{oed} = \sum_1^i \frac{a_v}{1 + e_0} \Delta\sigma'_z \, \Delta h_i \, , \tag{104}$$

$$a_v = -\frac{\Delta e}{\Delta\sigma'_z} \tag{105}$$

and further,

$$w_{oed} = \sum_{1}^{i} \Delta h_i \frac{C_c}{1 + e_0} \log \left(\frac{\sigma'_z + \Delta\sigma'_z}{\sigma'_z} \right) \tag{106}$$

(σ'_z – the original geostatic stress),

$$C_c = - \frac{\Delta e}{\Delta \log \sigma'_z} \tag{107}$$

(C_c – the compression index which is equal to a change in the void ratio for a logarithmic stress cycle). Unlike in the foregoing relations which are general, a special shape of the curve of oedometric compression is assumed in eqn. (106).

Since the compressibility factors m_v, a_v and C_c merely approximate the course of the oedometric compressibility curve and usually depend (sometimes with the exception of C_c) on the stress level, this curve can also be used directly in the determination of the settlement w_{oed}. The oedometric compression test does not reliably reproduce the in situ (geostatic) state of stress because the lateral pressures are merely a result of the transverse deformation of the sample; as pointed out by Hruban (1948), the horizontal stress is neither affected nor measured. Hence the compressibility is frequently too high.

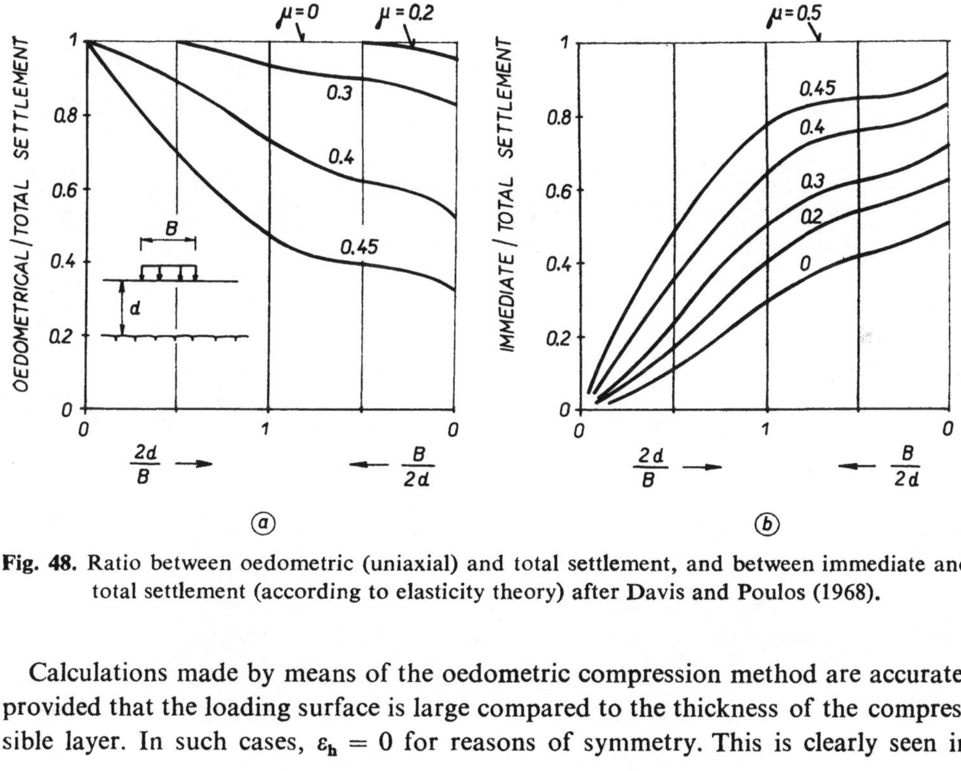

Fig. 48. Ratio between oedometric (uniaxial) and total settlement, and between immediate and total settlement (according to elasticity theory) after Davis and Poulos (1968).

Calculations made by means of the oedometric compression method are accurate, provided that the loading surface is large compared to the thickness of the compressible layer. In such cases, $\varepsilon_h = 0$ for reasons of symmetry. This is clearly seen in

Fig. 48a prepared according to Davis and Poulos (1968): if for a strip $B \rightarrow \infty$ or for a subsoil $d \rightarrow 0$ (d – the thickness of the compressible layer), then the settlement calculated by means of the oedometric compression method (the oedometric settlement) corresponds to the settlement w_T obtained by the application of the elasticity theory (i.e. to that including the lateral deformations of the subsoil). For a Poisson's ratio of the subsoil $\mu = 0.3$, the error is not greater than about 5% for $2d = B$.

The calculation procedure is well illustrated by Fig. 45a (for a two-phase halfspace, i.e. subsoil saturated with water): at the instant when a load is applied to the subsoil, the neutral stress (the path IG) increases by $\sigma'_{vF} - \sigma'_{vI} = \Delta\sigma'_z = \Delta u$. The method ignores the instantaneous settlement. The consolidation settlement proceeds along the path GE, the points I and E lie on the straight line of anisotropic consolidation K_0.

As Fig. 48a shows, for $\mu \neq 0$ the oedometric settlement w_{oed} is substantially smaller than the total settlement w_T. This result is not supported by experimental evidence. Thorough tests made by Burland (1969b, 1971) – Fig. 47 – show the settlement calculations by means of the oedometric compression method to be satisfactory up to a load equal to a half or a third of the limit load. This is explained by the measurements of the horizontal displacements beneath the corner of a model foundation strip shown in Fig. 47, which were about ten times smaller than the vertical displacements beneath the centre under a load close to the limit load, i.e. comparatively very small. This may be the reason why the oedometric compression method is frequently successful in practice (and is deemed to give sufficiently accurate results – Burland et al., 1977, p. 518, 536), even when the thickness of the compressible subsoil d is fairly large and when – according to Fig. 48a – the differences should amount to several tens of percent.

On the basis of the elasticity theory, one obtains for a uniformly loaded circular area (Burland and Wroth, 1974) the relation

$$\frac{w_{oed}}{w_T} = \frac{1 - 2\mu'}{(1 - \mu')^2} \tag{108}$$

If $\mu' < 0.24$, eqn. (108) gives $w_{oed} \geqq 0.9 w_T$ (for a likely range of μ' equal to 0.1 to 0.33 – see Burland and Wroth, 1974 – the range of w_{oed}/w_T is 0.99 to 0.75; if, however, $d/B < 1$, the error again decreases, for example, to only 10% and less for $\mu' \sim 0.3$ and $d/B \leqq 0.5$ – Giroud and Cordary, 1975). This degree of accuracy is satisfactory – see also Fig. 48a. Eqn. (108) can be used to correct w_{oed}; another possibility of correcting this quantity is provided by eqn. (98) and similar relations, noting that – unlike in the oedometer – $\varepsilon_v \neq \varepsilon_{vol}$ in foundation soils. The correction can be introduced directly in the basic equation (101) – Simons and Som, 1969.

Sometimes, however, the calculated settlements are higher than those observed. Besides other reasons (for example, disturbance of "undisturbed" samples) this may also be due to the deformation modulus increasing with depth, which fact is only imperfectly reflected in laboratory tests in the course of which the geostatic state of stress is not maintained.

A possible correction of this effect is the assumption that the thickness of the foundation soil (i.e. of the compressible subsoil) is finite. The compression of the foundation soil produced by a load then only propagates to a certain depth, the so-called active zone (see, e.g. Florin, 1961, p. 18 etc., Tsytovich, 1963, p. 364 etc.). Malikova (1972) states that the thickness of the active zone is less than a half of the foundation width in the case of large foundations, and smaller for sands than for clays. According to Konovalov's (1971) tests, this thickness is in direct proportion to the magnitude of the contact stress and to the foundation width, in indirect proportion to the deformation modulus of the foundation soil, and decreases with the depth of foundation (cf. Sections 3.2 and 3.10).

According to Steinbrenner (cf. Section 2.5.2), the elasticity theory, for example, Giroud's tables appended to this book, can be used for stress calculations made assuming the active zone to be e. g. of thickness d.

3.6 The Skempton-Bjerrum method

According to this method, the final settlement is the sum of the instantaneous settlement and the uniaxial consolidation. The method was proposed by Skempton and Bjerrum (1957) after Skempton et al. (1956) had drawn attention to the existence of the immediate (instantaneous) settlement. The principle of the method can be explained by referring to Fig. 45b. At the instant of a load increase, the neutral stress defined by eqn. (46) for $B = 1$ (the foundation soil is assumed to be completely saturated with water) is produced in the foundation soil. In this phase, the stress path is IB. Eqn. (46) can be given the form

$$\Delta u = \Delta\sigma_1[K_0 + A(1 - K_0)] \tag{109}$$

where $K_0 = \Delta\sigma_3/\Delta\sigma_1$. Since usually $0 < A < 1$ and $K_0 < 1$, one obtains $\Delta u < \Delta\sigma_1$; for $A = 1$, $\Delta u = \Delta\sigma_1$ and the consolidation settlement is the same as that during uniaxial compression.

When a load is applied to the foundation soil, the instantaneous settlement takes place. This settlement is determined either by calculation using the theory of linear halfspace ($\mu = 0.5$; E is usually the tangent deformation modulus at the beginning of the curve describing the $(\Delta\sigma_v - \Delta\sigma_h) - \varepsilon_v$ relation for undrained triaxial test) or directly from a triaxial test by means of the relation

$$w = \int_0^d \frac{\Delta\sigma_v - \Delta\sigma_h}{E}\, dz \tag{110}$$

(d — the depth of the compressible subsoil). A thorough analysis of the problems concerning the calculations of the instantaneous settlement can be found in the works of D'Appolonia and Lambe (1970) and D' Appolonia et al. (1971)*). Fig. 48b shows the ratio of the immediate settlement to the total settlement (after Davis and Poulos): if $\mu = 0.5$, the total settlement is immediate; for $d \to 0$ or $B \to \infty$, the immediate settlement becomes zero.

The consolidation phase assumes the consolidation to be uniaxial along the stress path *CE*. Hence the stress path consists of two segments, *IB* and *CE*, and is discontinuous; the solution is approximate (Lambe, 1964).

Compared with the oedometric compression method, the Skempton-Bjerrum method yields smaller settlements for small values of A, i.e. for overconsolidated clays; for $A = 1$, on the other hand, the settlement is increased by the instantaneous settlement (see, e.g. Lambe, 1963).

For $A < 1$, the consolidation settlement is smaller by a factor α_0 than the settlement obtained by means of the oedometric compression method. Skempton and Bjerrum (1957) present diagrams of the factor α_0 as it depends on the magnitude of A, the shape of the foundation (a strip or a circle), and the ratio of the depth of the compressible subsoil to the width of the foundation. Critical comments on their method and corrections of their diagrams are offered by Scott (1963, p. 281 et seq.).

3.7 Empirical methods

These methods start from the result of some type of field test of the mechanical properties of the foundation soil, which cannot uniquely be interpreted theoretically, and use empirical correlations to predict the settlement of an actual foundation (usually on a sandy subsoil). Their successful application depends to a considerable degree on experience. The following tests are used: the plate loading tests (Terzaghi and Peck, 1967, p. 488 et seq.; Hobbs, 1974); the standard penetration tests (Terzaghi and Peck, 1967, p. 491 etc.; Meyerhof, 1965; Sutherland, 1974); the static penetration tests (De Beer, 1967; Schmertmann, 1970); the pressiometric tests (Ménard and Rousseau, 1962; Ménard, 1971), etc. A systematic review of the penetration tests was prepared by Sanglerat (1972).

As an example of the applicable relations, one can consider a relation proposed by Terzaghi and Peck (1967) for the determination of the settlement of a foundation on sandy subsoil, viz.

$$w = w_0 \frac{4}{\left(1 + \frac{B_0}{B}\right)^2} \tag{111}$$

---

*) In this study, the authors introduce a correction for the effect of non-linearity of the load — initial (instantaneous) settlement relation; this non-linearity depends on the geostatic state of stress in the subsoil and on the load level.

where w, B – the settlement and the smallest dimension of the foundation; w_0, B_0 – the settlement and the smallest dimension of the loading plate ($B_0 = 0.3$ m). According to measurements, the reliability of eqn. (111) is low – predictions based on it are rather conservative because actual settlements are apt to be higher by a factor of up to 5 (Bjerrum and Eggestad, 1963; Sutherland, 1974). Eqn. (111) is oversimplified and fails to describe a number of factors controlling the settlement. Carrier and Christian (1973) showed the settlement to be a logarithmic function of B.

On the basis of a previous empirical relation of Terzaghi and Peck, relating the permissible load and the foundation width to the standard dynamic penetration strength obtained by the SPT method, and eqn. (111), Meyerhof (1965) proposed for sands the following relation between the settlement and the standard penetration strength N (after conversion to metric units)*)

$$w \doteq \frac{2q}{N} \qquad \text{for} \quad B \leqq 1.2 \text{ m}$$

$$w \doteq \frac{3q}{N}\left(\frac{B}{0.3 + B}\right)^2 \qquad \text{for} \quad B > 1.2 \text{ m} \tag{112}$$

In the above, q is the load acting on the foundation (MPa), B – the width of the foundation (m), w – the settlement (cm). As subsequent studies have shown, before being substituted in eqns. (112), the standard penetration strength must be corrected, especially as to the influençe of the effective pressure of the overburden, for even when the relative density of sand remains the same, the value of N increases in proportion to this pressure. Such a correction was proposed by Peck and Bazaraa (1969). A slightly different procedure was adopted by Peck, Hanson and Thornburn (1974) who revised the original equation of Terzaghi and Peck, which relates the permissible load, the width of the foundation and the value N, and differentiated it for various depths of foundation.

Two objections can be put forward against the empirical procedures which make use of eqns. (111) and (112). On the one hand, they are not empirical enough – they do not contain a free parameter which would describe the regional conditions, i.e. the geological origin, the degree of overconsolidation, etc. On the other hand, they are very simple – they fail to take into account the effect of the shape of foundations (various boundary conditions are speculatively assumed to compensate one another – Terzaghi and Peck, 1967, p. 489), the effect of the subsoil thickness, etc., which are incorporated in the theoretical formulae.

The second objection constitutes a reason for the development of a procedure in which the results of penetration (or pressiometric) tests are used only for the deter-

*) N is an average value for the subsoil between the base of foundation and a depth equal to the width of the foundation (Terzaghi and Peck, 1967, p. 490).

mination of the deformation modulus and the settlement is calculated from equations based on the elasticity theory or the theory of uniaxial compression (with certain modifications, if necessary). This procedure forms the basis of all the empirical methods which will be discussed later in this section. The only difference between these various methods is the technique (i.e. the dynamic or the static penetration test, or a pressuremeter) adopted for the determination of the deformation modulus.

An empirical approach formulated in the above way is an unquestionable advance, for it enables the analysis to be better differentiated (De Beer, 1965; D'Appolonia et al., 1970), while retaining its empirical nature thanks to the correlations between the data of complex field tests and the deformation modulus*). Since the results of such field tests, besides being affected by the subsoil compressibility, depend on numerous additional parameters, these correlations can only be roughly approximate. This is the reason why some authors reject the procedure in toto (Terzaghi and Peck, 1967, p. 492) or oppose all attempts to refine it (De Beer, 1965).

For static cone penetration in sands, De Beer (1965) gives the equation

$$w = \int_0^{\infty} \frac{1}{C} 2.3 \log\left(\frac{\sigma_z + \Delta\sigma_z}{\sigma_z}\right) dz \tag{113}$$

which is analogous to eqn. (106) (σ_z – the vertical geostatic pressure). The compressibility coefficient C is:

(a) for normally consolidated sand

$$C = \frac{3}{2} \frac{C_p}{\sigma_z} \tag{114}$$

(b) for overconsolidated sand

$$C = \frac{3}{2} \frac{C_p}{\sigma_z} \frac{A_0}{C_0} \tag{115}$$

In keeping with eqn. (42)

$$C\sigma_z = \frac{E}{1 - \dfrac{2\mu^2}{1 - \mu}} \tag{116}$$

In eqns. (114) and (115), C_p denotes the in situ cone resistance of sand; C_0, A_0 are the coefficients of oedometric compressibility of a sample having the same density as in the in situ condition, the first coefficient corresponding to the branch of increasing loading, the second to that of unloading. The form of eqns. (114) and (115) was

*) Such correlations are deduced either from the measured settlements of structures (D'Appolonia et al., 1970) or from plate loading tests (Schmertmann, 1970).

proposed by Buisman. At present, the constant coefficient 1.5 appearing in these equations, is replaced by a coefficient which ranges from 0.4 to 8, depending on the nature of the subsoil (Sanglerat, 1972, pp. 346 and 356).

Schmertmann (1970) chooses a different procedure for calculating the settlement from the results of a static penetration test. He starts from eqn. (91), replaces the exact course of $\varepsilon_z(z)$ by a bilinear (a triangle with its apex at a depth $B/2$ or $D/2$), and assumes that $\varepsilon_z = 0$ for z/B $(z/D) \geqq 2$ (i.e. the depth of the active zone is equal to $2B$ or $2D$). He then obtains the deformation modulus E of sand on the basis of its correlation with C_p, viz.

$$E = 2C_p \tag{117}$$

In contrast to De Beer, Schmertmann does not assume that the stress in the subsoil follows the K_0 path.

D'Appolonia et al. (1970) calculate the settlement from the results of standard penetration tests using eqn. (89), and assume that $\mu = 0.25$ for sand. They use the following direct empirical correlation (E in MPa):

(a) for overconsolidated sand:

$$E = 53 + 1.32N \tag{118}$$

(b) for normally consolidated sand:

$$E = 21.2 + 1.04N \tag{119}$$

Once the correlation between C_p and N is known, any method based on the static or the dynamic penetration can be used to calculate the settlement. The correlations used are usually of a simple form, viz.

$$C_p = kN \tag{120}$$

where the coefficient k ranges from 2.5 to 16 – it increases with the grain size of cohesionless soils (for comprehensive information, see Sherif, 1973 p. 79; Sutherland, 1974), C_p in MPa/10.

A correlation which is much more complicated than eqn. (120) was derived by Sherif (1973, p. 169; see also Schultze and Sherif, 1973). According to it, the oedometric deformation modulus ($= C\sigma_z$ by eqns. (114), (115), (116)) depends not only on the penetration resistance N but also on the width and the depth of foundation as well as on the kind of soil being considered.

A pressuremeter measures the dependence between the radial stress acting on the walls of a vertical borehole, and the changes in the diameter of the borehole (see, e.g., Caquot and Kérisel, 1966, p. 101 et seq.; Prévost and Hoëg, 1975). The linear part of this dependence defines the so-called pressiometric deformation modulus E_p (cf. Gibson and Anderson, 1961). The equation for calculating the settlement with its aid (Ménard, 1971) is as follows:

$$w = \frac{qB}{E_{oed}}\frac{\lambda_1}{9} + \frac{qB_0}{E_p}\frac{1}{3}\left(\frac{B}{2B_0}\lambda_2\right)^{\alpha_0} \tag{121*)}$$

(for $B > 2B_0$) where $B_0 = 0.3$ m; B is the width or the diameter of the foundation; λ_1, λ_2 are the shape factors of the foundation, E_{oed} is the oedometric deformation modulus, $E_{oed} = \Delta\sigma_z/\varepsilon_z$ defined by eqns. (42) and (116). According to Ménard (1971) who calls it the coefficient of structure, the exponent α_0 has the following values: 1 (elasticity theory); 2/3 (clay), 1/2 (loam), 1/3 (sand, gravel), 1/4 (boulders). It expresses the fact that the settlement does not usually increase in direct proportion to the width of the foundation. Ménard (1971) defines the exponent α_0 by the equation

$$\frac{G}{\alpha_0} = \frac{E}{2(1 + \mu)} \tag{122}$$

which, for $\alpha_0 = 1$, changes to a relation familiar from the elasticity theory. The magnitude of α_0 thus characterizes the deviation of a medium from isotropy (cf. Poirot, 1971).

Eqn. (121) contains two terms. According to Ménard (1971), the first term on the right-hand side describes the effect of the spherical stress tensor, the second term, the effect of the stress deviator on the deformation of the subsoil. The first term predominates in the case when the thickness of the compressible subsoil is small compared with the dimensions of the foundation; the second in the case when this thickness is large.

Measurements made by a number of authors show the existence of a correlation between E_p and C_p which is of the type

$$E_p = kC_p \tag{123}$$

where $k = 2$ to 3.7 for cohesive soils, and $k \leqq 1$ for sands (Sanglerat, 1972, p. 313 et seq.).

Other empirical relations were proposed by Kondner (1962) – a hyperbolic relation between the load and the settlement of a group of floating piles in cohesive foundation soil, by Bond (1961), Chin (1969), etc.

Empirical relations are derived for a definite set of experimental data. Their shortcoming is the uncertainty of when one can reliably extrapolate beyond the limits of this set.

In conclusion, it seems to the point to state – in harmony with Lambe (1973) – that many methods of settlement prediction adopt an engineering approach in which

*) Bru et al. (1973) reproduce this equation in a slightly modified form: the first term on the right-hand side is multiplied by the parameter α_0, the second is increased by 1/3.

the method of calculation is closely related to the method of selecting the physico-mechanical parameters of the subsoil. Judged by this fact, many methods of settlement calculations are found to be of a semi-empirical character.

3.8 The finite element and allied methods

Automatic computers can be applied to settlement calculations carried out in accordance with the classical methods: the relevant formula is the algorithm of the solution (Kandaurov et al., 1969; Schiffman et al., 1970; Harrison et al., 1972). Special attention should, however, be accorded to numerical methods whose advancement was enabled by the advent of computers. The iterative or incremental analysis of non-linear problems which is usually involved makes use of the mechanism of back coupling and requires, therefore, a great number of numerical operations. The principal advantage of these methods and the reason for their successful application to settlement calculations is the fact that the stress-strain relation of the mathematically modelled medium follows the experimental non-linear stress-strain relation with a required degree of accuracy, and that exact boundary conditions are introduced (Section 2.9).

These methods eliminate the mathematical incorrectness of classical methods (Sections 3.3 to 3.6) which employ one stress-strain relation for stress calculations and another for deformation calculations. They make it possible to calculate the load-settlement relation within the whole range of loads, regardless of the magnitude of the load, i.e. up to the limit load. Thus they do away with the artificially formed division between deformation (settlement) and stability (limit load) problems. They do so at the price of the loss of mathematical elegance and accuracy of analytical solutions. In contrast to the latter, numerical solutions are usually presented for a finite number of points of the subsoil, selected in the course of the process of discretization of the medium.

The limitations of these methods are mostly due to experimental data. An experimental stress-strain relation can only be determined under special conditions of loading (state of stress) and depends – as generally acknowledged – on the stress path (review in Feda, 1972). At present no one knows how great is the sensitivity of various soils to various stress paths. Other effects which come into play have to do with sampling of undisturbed samples (even so-called perfect samples change their state of stress during sampling), some uncertainty as regards the geostatic (and geodynamic) state of stress, etc.

Some inaccuracy may be caused by the fact that – unlike an actual subsoil – a mathematical model of the subsoil possesses a rigid boundary at a distance from the foundation. The calculated settlement can then differ from the actual settlement by 10% to 15% (Carrier and Christian, 1973; Simons and Rodrigues, 1975); this inaccuracy can, however, be corrected by the application of a correction factor (Carrier and Christian, 1973).

A study of Hoëg et al. (1968) does not tackle the problem of a true description of the stress-strain relation of real soils. It assumes an elasto-plastic (ideally elastic and ideally plastic) stress-strain relation and Tresca's criterion of failure. It employs a specific mathematical model (lumped-parameter method) of the medium: the foundation soil is divided (by means of a square net) into mass points (at which the forces and the displacements are applied) and into stress points (at which the stresses and the strains are at work). The solution is incremental: for each increment of load, the increment of strain is calculated from the displacement of the mass points, and from this increment of strain, the increment of stress. After each increment of load, a correction is made to satisfy Tresca's condition. The mathematical model approximates the differential relations by means of the finite difference method. Because of the idealized constitutive relations, the solution of the problem of settlement is merely qualitative, accomplished by an analysis of the effects of various factors, including that of local plastic regions (some results are also presented by Lambe and Whitman, 1969, p. 195 et seq.). A similar procedure was adopted by Biarez and Boulon (1971) using the von Mises criterion, and by Giroud et al. (1970).

The possibility of determining the extent of local plastic regions and the state of stress in the case when these regions are present is a point by which this method excels the finite element method in its quasi-elastic formulation.

The finite difference method was also used by Kryzhanovskiy et al. (1975) for solving the stress-strain relation of a model foundation on sandy subsoil. Fig. 49 shows the load-settlement curve for constitutive equations which are realistic to various degrees. Their parameters were measured in an apparatus using a cubic sample and three independently variable principal stresses. Fig. 49 shows a clearly marked effect of dilatancy which gradually increases with the magnitude of the load: a 10 cm settlement which, without the effect of dilatancy, is obtained at a load of 1.4 MPa, is produced by the application of up to double this value — 2.8 MPa under the effect of dilatancy (the curves *3* and *1*). Although the constitutive relation is non-linear, the settlement w remains linear for a considerable range of q; this fact demonstrates that one can hardly form a unique opinion as regard the linearity of a constitutive relation of the subsoil on the basis of measured linearity of the $q-w$ relation. The calculated settlement was about 12% higher than the measured one.

As the solution also shows, the originally homogeneous subsoil becomes deformationally non-homogeneous in the course of loading, and plastic regions grow gradually larger beneath the foundation, in the neighbourhood of the foundation centreline (see also Lambe and Whitman, 1969, p. 169); this fact suggests local shear failure of the foundation soil.

The finite element method is used in soil mechanics at an ever increasing rate (cf. Radhakrishnan and Reese, 1970; Desai and Abel, 1972). As mentioned earlier, its main problem is the parametric expression of a non-linear stress-strain relation.

A great advantage of the finite element method is that it can be applied to solutions of continuous as well as discontinuous media, is capable of managing complicated constitutive relations, even those which are spatially variable (different relations for different elements) and permits the introduction of realistic boundary conditions.

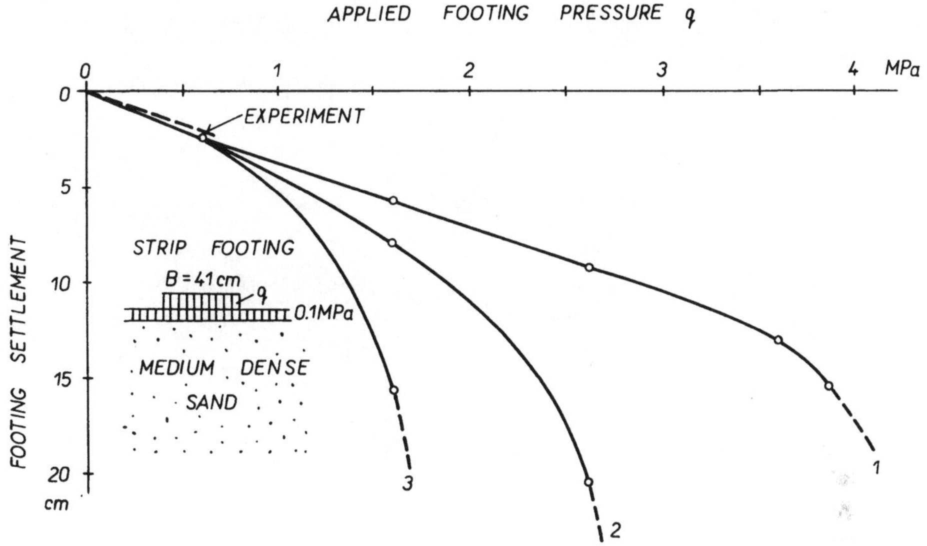

Fig. 49. Relation between load and settlement of a model foundation strip on sandy subsoil, calculated by the method of finite differences (Kryzhanovskiy et al., 1975); *1* — dilatancy and other real deformation properties included in the calculation; *2* — effect of dilatancy restricted; *3* — effect of dilatancy excluded.

The usual formulation of the finite element method is the so-called quasi-elastic formulation. The stress-strain relation is defined by means of the deformation modulus E and Poisson's ratio μ which are both dependent on the magnitude of stress. The general variational formulation of the finite element method requires that the solution should lead to minimization of a certain functional of energy character (the total potential energy of a medium which is assumed to be perfectly elastic; in soils under loads once applied, only the deformation energy is involved instead of the total potential energy — cf. Kolář et al., 1971; Desai and Abel, 1972, etc.). However, the finite element method need not start from the variational formulation and can also be used in cases when no variational functional exists (Zienkiewicz, 1970).

In the calculation the subsoil in question is conveniently divided into elements in which one estimates the deformation parameters E, μ and calculates the stresses and strains. Since the deformation parameters of each element depend on the element's stress and the stress in turn depends on the deformation parameters (the stiffness matrix of the elements and of the system), the analysis must be repeated over and over for the changed parameters in order that the stress-strain relation of each element should be compatible with the postulated non-linear law. Besides the method of quasi-elastic parameters which vary with each iteration step, one can also use a method in which plastic strains are introduced as the initial deformations of the elements, and the quasi-elastic parameters are constant (Zienkiewicz and Cheung, 1967; Malina, 1969) or the method of initial stresses. In this way a distinction can be made between the "stiffness perturbation method" and the "load perturbation method" (Goodman and Dubois, 1972).

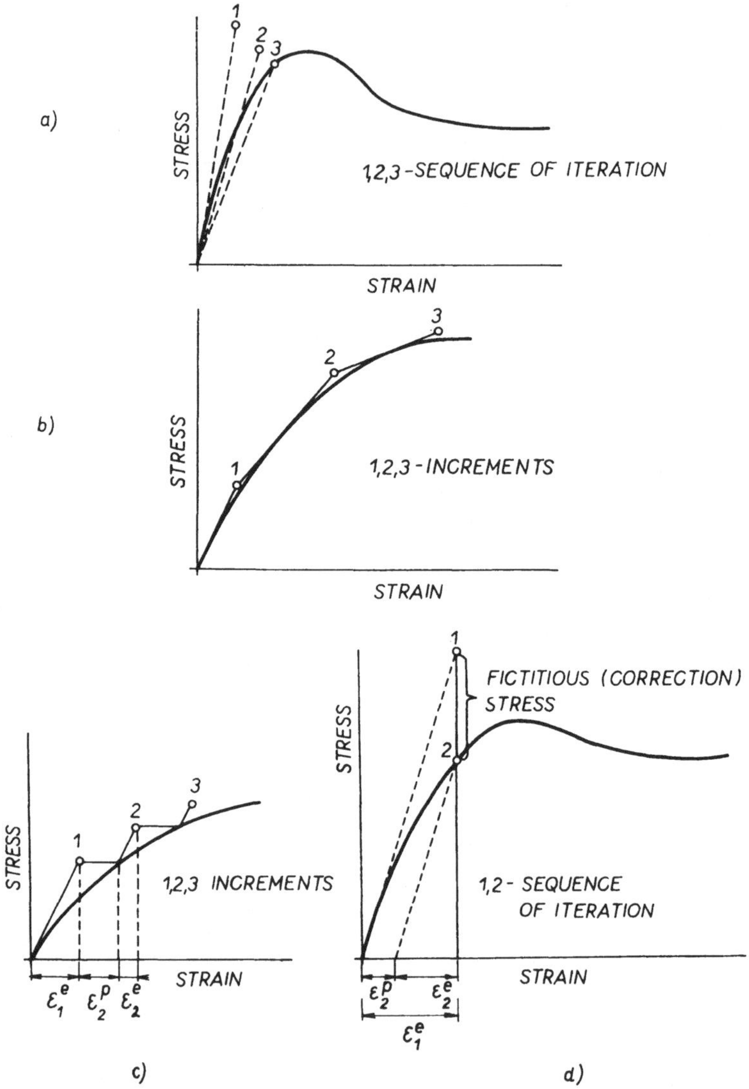

Fig. 50. Characteristics of non-linear stress-strain relation for the iterative, incremental and initial plastic strain methods.

The non-linear stress-strain relation of the method of quasi-elastic parameters can be approximated in two ways: iterative and incremental (Duncan and Chang, 1970).

In the first case the same load is applied several times in succession. For each step the stress and strain in each element are calculated and the original deformation parameters are corrected according to the result; this is repeated until the stress and strain in each element correspond to the experimental non-linear relation. Fig. 50a shows the iteration steps 1 to 3 – the last one is already satisfactory.

In the second case the change of load is divided into a number of increments. The stress and strain obtained in each element for the first increment are used in the determination of the corrected values of the deformation parameters for the next increment. Thus the actual stress-strain relation represented by a smooth curve is replaced by a polygon (Fig. 50b). This procedure can be combined with the first, the iterative procedure (mixed procedure – see, e.g., Desai and Abel, 1972, p. 219 et seq.).

The iterative procedure is advantageous in that it can also be used in cases when the stress-strain relation is unstable (i.e. when a post-peak decrease in stress occurs). In such cases the incremental method would have to introduce a negative deformation modulus which cannot be done in the finite element method; neither can $\mu \geqq 0.5$ be used (since the medium involved is incompressible, the stiffnesses of the individual elements are infinite for $\mu = 0.5$). An advantage of the incremental method is the fact that the effect of the initial (geostatic) state of stress in the halfspace can be introduced in an easy manner, and that the stress and strain are calculated for each load increment.

The method of initial plastic deformation is shown schematically in Fig. 50c. Of interest is its alternative shown in Fig. 50d. Should the state of stress of an element obtained in the first step be larger than that specified by the limiting condition (e.g. Mohr's circle – the procedure can, however, be generalized), that particular element is unloaded (the reversible strain ε_1^e being produced during the process), and the plastic strain ε_2^p assigned to it in the second step is chosen so as to make, together with the deformation of the element with the reduced strain ε_2^e, the total strain of the element equal to that obtained in the first step (i.e. to ε_1^e). The reduced strain satisfies the a priori non-linear stress-strain relation (point *2*) and is determined using the same quasi-elastic parameters (i.e. the stiffness matrix of the elements and of the system has undergone no change, the load perturbation procedure). The reduction of the state of stress in the element disturbs the equilibrium between the external and the internal forces. The equilibrium must be restored by the addition of "fictitious" internal forces to the nodes of the element; these forces produce a corrective field of stress and thus change the states of stress in the remaining elements. This is the way in which a natural transfer of stresses from overloaded elements to elements subjected to lesser loads is effected.

According to Malina (1959, pp. 20 and 26), the method of initial plastic deformation makes it possible to introduce the volume changes of the medium (dilatancy and contractancy) produced during shear. Malina, however, does not introduce them in his work, and is satisfied to consider elastic behaviour (zero volume changes during shear).

A simple bi-linear constitutive relation in combination with the finite element method was used for settlement calculations by D'Appolonia and Lambe (1970).

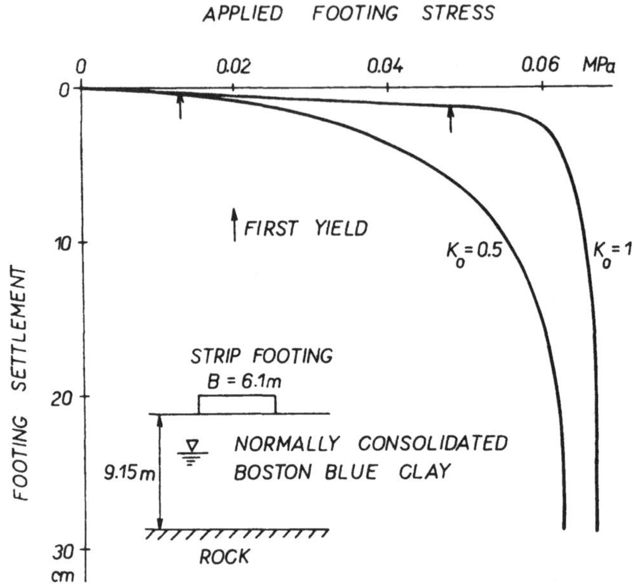

Fig. 51. Effect of geostatic stress anisotropy on the settlement of a foundation strip (D'Appolonia and Lambe, 1970).

One of the useful results obtained by these authors – the effect of the geostatic state of stress on the settlement – is shown in Fig. 51. For a safety factor of 2 to 3, the settlement is by 150% – 50% larger for $K_0 = 0.5$ than for $K_0 = 1$. As indicated by the arrows, the plastic regions start to develop much sooner in the case of $K_0 = 0.5$ (at a safety factor of 5) than of $K_0 = 1$ (at a safety factor of 1.4).

A study published by Girijavallabhan and Reese (1968) can serve as an example of the iterative procedure. The authors solved an axi-symmetrical problem, i.e. a circular foundation on the surface of soft clay of finite thickness. They assumed that $\mu = 0.48 =$ const. For a chosen deformation modulus they calculated the shear strain at the centre of the elements, γ_{oct}, determined G from an experimental relation between the shear modulus and γ_{oct} (the relation was derived on the basis of a triaxial test with clay used in the model loading tests, and was stable); from G thus obtained, they established a new, corrected value of E, etc. and continued this process until an agreement was reached between the calculated and the experimental stress-strain relations. As Fig. 52 shows, the results of the calculation agree very well with those of the model loading test.

The incremental method was applied to the calculation of the load-settlement relation by Duncan and Chang (1970). These authors solved the variations of the load and the settlement of a rectangular foundation with a ratio of sides of about 1 : 5 on sand as a problem of plane deformation. They used triaxial tests at a constant cell pressure to obtain the characteristic of the non-linearity of their stress-strain relation. The two errors which they committed by proceeding in this way compensated one another in a sense.

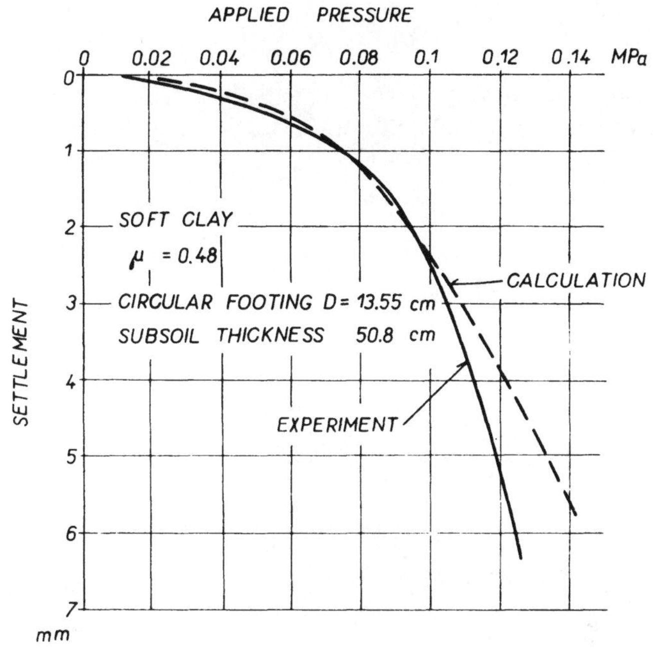

Fig. 52. Load-settlement curves of a circular foundation on clay—model experiment and calculation by the finite element method (Girijavalabhan and Reese, 1968).

Although they experimentally established that Poisson's ratio (or more exactly, the coefficient of transverse extension) varied in the range 0.11 to 0.65 in proportion to the stress level, they assumed that $\mu = 0.35 =$ const. in order to simplify the calculation.

When using the incremental analysis, one must determine the magnitude of the tangent deformation modulus E_t as it depends on the state of stress and strain in each element. Duncan and Chang (1970) accept a hyperbolic stress-strain relation for the triaxial test, and the following parabolic relation between the initial tangent modulus E_i (which is one of the parameters of their stress-strain relation) and the magnitude of the effective cell pressure (cf. eqn. 82):

$$E_i = ap_a\left(\frac{\sigma_3'}{p_a}\right)^n \tag{124}$$

where p_a is the atmospheric pressure (in the units of σ_3' and E_i); a, n are a coefficient and an exponent, which are parameters of this relation (n varies from 0.4 for gravel and rockfill, to 1 for saturated clay and undisturbed rock; the values of n were tabulated by Kulhawy – cf. Mitchell, 1969). The resulting equation is of the form

$$E_t = \left[1 - \frac{R_f(1 - \sin \Phi_f')(\sigma_1 - \sigma_3)}{2c' \cos \Phi_f' + 2\sigma_3' \sin \Phi_f'}\right]^2 ap_a\left(\frac{\sigma_3'}{p_a}\right)^n \tag{125}$$

In the above, c', Φ_f' – the shear parameters; R_f – the ratio of the actual to the asymptotic difference of principal (axial and radial) stresses in a triaxial test.

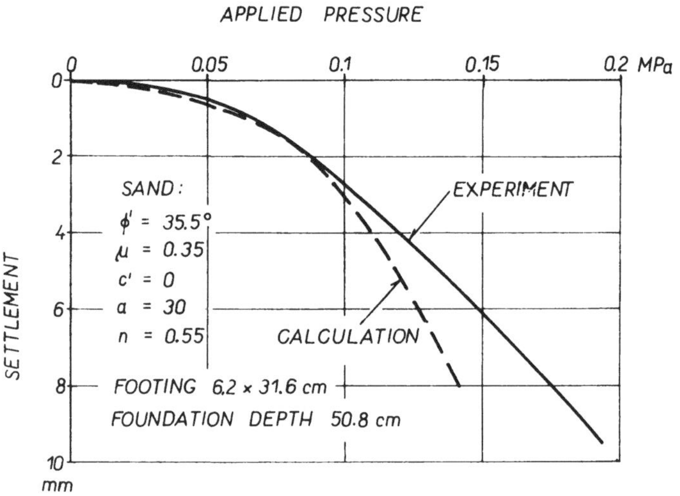

Fig. 53. Load-settlement curves of a rectangular foundation on sand – model experiment and calculation by the finite element method (Duncan and Chang, 1970).

A comparison between the values obtained by calculations of Duncan and Chang (1970) and the results of model loading tests is shown in Fig. 53 (which also gives the parameters of eqn. (125)). For low loads, the agreement is very good; for high loads, it could be improved – according to the opinion of the authors – by using higher values of the angle of internal friction and of Poisson's ratio μ, which are probably higher in this phase anyway. Mitchell (1969) presents an example of the same calculations, relating to the settlement of two square foundations (with a side of 0.45 m and 1.5 m, respectively) on a non-homogeneous (layered) subsoil, and also obtains a good agreement with experiments. Desai (Desai and Abel, 1972, p. 331 et seq.) compares the settlement of a rigid circular model foundation made of steel (7.6 cm diameter, 1.27 cm thick) on a two-layer clayey subsoil, with that calculated by means of the finite element method (the constitutive relation was derived on the basis of undrained triaxial tests at different cell pressures). The agreement between the two sets of values is very good.

It can be stated in conclusion that – despite obviously inaccurate initial assumptions (a three-dimensional problem is solved as a plane problem, the mechanical parameters are derived from standard triaxial tests, Poisson's ratio is chosen to be constant) and physically poor description of the subsoil by the quasi-elastic method (the method starts from the assumption of elastic behaviour of soils: it pays no heed, for example, to volume changes produced by shear, i.e. to dilatancy and contractancy) – analyses made by means of the finite element method are qualitatively satisfactory and quantitatively reliable enough for loads for which the settlement is usually calculated. The discord between calculations and reality (Fig. 52 and 53) can probably be explained by local subsoil failure in the first case and by the variability of Poisson's ratio with load in the second case. Since this ratio increases with increasing load, the actual settlement slightly decreases, according to the elasticity theory and Fig. 49 (cf. Fig. 53 for loads higher than 0.1 MPa).

A variable Poisson's ratio can be introduced just as a variable deformation modulus. If the relation between ε_a and ε_r is assumed to vary according to a hyperbolic law, then (Desai and Abel, 1972, p. 321; Kulhawy and Duncan, 1972)

$$\mu_t = \frac{g - f \log\left(\dfrac{\sigma_3'}{p_a}\right)}{(1 - d\varepsilon_a)^2} \tag{126}$$

where

$$\varepsilon_a = \frac{\sigma_1 - \sigma_3}{a p_a \left(\dfrac{\sigma_3'}{p_a}\right)^n \left[1 - \dfrac{R_f(\sigma_1 - \sigma_3)(1 - \sin \Phi_f')}{2c' \cos \Phi_f' + 2\sigma_3' \sin \Phi_f'}\right]} \tag{127}$$

In eqns. (126) and (127), d, f, g are parameters; the remaining symbols have the same meaning as in eqns. (124), (125); μ_t – the magnitude of the tangent Poisson's ratio.

The good agreement which was found to exist between the calculated values (obtained by means of the quasi-elastic methods) and experiments in all the examples quoted above can be attributed to mutual compensation of errors rather than to the accuracy of the method. Just as any new method, the finite element method requires to be verified by case histories (Duncan, 1973) which are only inadequately replaced by model laboratory tests. Since the measured and the calculated strains are only examined at a single point, a comparison of the measured settlement with the calculated one is unsatisfactory. To demonstrate that the mathematical model describes the physical reality*), one must compare the deformations of the surface as well as the displacements of various points of the subsoil. Detailed analyses of the behaviour of actual structures, such as those presented by Burland et al. (1973) – see Section 3.10 – are particularly valuable in this respect. The finite element method has been found to work successfully also in such analyses.

The constitutive relations with two parameters – E and μ – which are available at present, are not appropriate for subsoils (cf. Section 2.8.2.3), and it seems more suitable to start from the general non-linear theory of the stress and strain tensor relation. This procedure was chosen by Wibel (1971). Its application made it necessary to determine twelve mechanical parameters for the pre-peak branch of the stress-strain diagram, and twelve other parameters for the post-peak branch. These parameters were obtained for sand on the basis of a standard triaxial test (including the measurement of transverse deformations of the samples; friction on the bases of the samples was reduced) and of an oedometric test. Wibel calculated the settlement of a circular foundation (D = 25 cm) on the surface of a sandy subsoil, and compared it with experiments. The qualitative agreement was very good but, just as in Fig. 53, the measured settlement was smaller than the calculated one. Wibel (1971) explains this – satisfactorily on the whole – by the fact that only two types of instruments with special boundary conditions were used to determine the mechanical parameters. This should be understood to mean that at a general point of the subsoil the stress paths do not resemble the stress paths applied in laboratory tests.

An analytic expression more reliable than that provided by a hyperbolic function can be obtained by means of spline functions. These functions interpolate the stress-strain relation between two nodes so as to yield a smooth transition from one interval to the next. Desai (1971) compares the calculated and the measured load-settlement relations for a model circular foundation on a cohesive subsoil. He obtains results which are qualitatively similar to those shown in Fig. 52; the approximation of the stress-strain relation by means of a spline function yields better results than an hyperbolic relation.

Since not all types of stresses can be simulated in test apparatuses, the analytic procedure described above has certain limitations. The stress paths as well as the stress levels are different at different points of the subsoil. Hence a realistic theoretical interpretation of the mechanical behaviour of soils, which would enable the laboratory results to be extrapolated, is demanded as necessary.

An example of such a conception is the state boundary surface (Roscoe and Burland, 1968). The possibility of using the theory of the state boundary surface in the finite element method (because the corresponding constitutive relation is highly complicated, numerical methods must be adopted) was studied by Zienkiewicz and

*) As mentioned in Section 3.2 and in connection with Fig. 49, linearity or non-linearity of the load-settlement relation does not make it possible to characterize uniquely the physical nature of the subsoil.

Naylor (1971). They showed that this theory is a special case of an explicit relation between the stress increment and the strain increment which is general for various incremental theories of plasticity, and that this explicit relation can directly be used in the finite element method.

The state boundary surface theory is one of the possible formulations of the so-called state equations of soils (Feda, 1972). In combination with the finite element method, the incremental constitutive relations which describe the dilatational behaviour of soils, are clearly the most suitable of all for settlement predictions. A proper choice of the site and of the direction of sampling must ensure a realistic description of possible non-homogeneity and anisotropy of the subsoil.

The numerical methods described so far are applicable to the solution of stresses and strains in continua (continuous media).

As already mentioned in Sections 1.2.3 and 2.2, the finite element method can equally well be adopted to the solution of stresses and strains in discontinuous media.

The elements which adjoin the discontinuity surface, have their nodal points on this surface. Essentially, these elements are not different from the elements of continuous media. The discontinuity surface is characterized by special joint elements which are of zero thickness, and by a specific constitutive relation which contains the so-called unit stiffnesses across (normal) and along (shearing) the joint, of these elements (Goodman et al., 1968). The surfaces of physical discontinuity can at the same time be the surfaces of mathematical discontinuity of the field of stresses and strains, depending on the boundary conditions of the problem.

The closeness of a mathematical model thus constructed to the physical nature of the medium further enhances the possibility of prescribing the dilatational behaviour of the discontinuity surfaces (Goodman and Dubois, 1972; Ghaboussi et al., 1973). In this way one can describe the couplings between the stress deviator and the spherical strain tensor in a given medium, which fact must be regarded as a remarkable advance over the quasi-elastic solutions.

The application of this procedure to the calculations of the deformations of a subsoil loaded by a foundation, was demonstrated by Ghaboussi et al. (1973). Since, for example, the joint elements successfully model the contact between a rigid foundation and a subsoil, they serve well in the description of the effect of the contact stress on the distribution of σ_z at various depths beneath a rigid foundation.

3.9 Notes on the choice of mechanical parameters of subsoils

Essentially, the choice of mechanical parameters of subsoils can be made in two ways: directly (by measuring the mechanical parameters which are needed for calculating the settlement of the subsoil in question), and indirectly.

The following three methods are available for indirect determination of the deformation parameters of a subsoil:

(1) The deformation parameters are derived from the measurements of other mechanical properties of the subsoil, using a relation, for example, of the type

$$E_u = kc_u \tag{128}$$

where E_u is the deformation modulus and c_u the strength in undrained test, test with quick loading (e.g. for the calculation of the immediate settlement), etc. of a water-saturated subsoil; k is a coefficient. The magnitude of this coefficient was determined by many authors. A summary of their results shows (Simons, 1974; Butler, 1974) that k = 40 to 3000 for normally consolidated and slightly overconsolidated clays, k = 140 to 800 for overconsolidated clays. It is seen that a correct choice of k is necessarily based on experience, considerations concerning the magnitude of the load, etc.

Other relations that can also be used for the purpose on hand are those between the modulus of compressibility and the static penetration strength (eqn. (114)), between the deformation modulus and the limit pressiometric pressure (Van Wambeke, 1975), etc.

(2) A deformation parameter (usually the modulus of oedometric compressibility) is obtained from the measurements of another physical property of the subsoil (void ratio, moisture content, liquid limit) to which it is closely related. Many correlations of this kind have been published (Skempton, 1944; Cozzolino, 1961; Dzwilewski and Richards, 1974; Simons, 1974, and others). Since compressibility depends on more than a single physical property, they frequently differ from one another to a considerable degree.

(3) The deformation parameters (usually E and μ) are tabulated for various types of subsoils classified according to their index properties. The tabulated values are a result of a combination of direct measurements, experience and calculations made on the basis of measured settlements, and are of regional validity. Although it is usually semi-empirical since it a priori assumes a certain mathematical model of the subsoil, the last method – calculation of the deformation modulus from measured settlements – is of particular value. The tables of deformation parameters are included in many Standards (for example, Czechoslovak, Polish, Soviet), and can gradually be refined on the basis of settlement measurements (see, for example, Boleński, 1973).

Indirect methods are of unequal usefulness and frequently quite subjective. They are usually of regional validity and of more or less empirical character. This is why direct measurements are more suitable from the theoretical point of view. A critical review of their recent advances is presented by Ladd et al. (1977).

The method which is to be used for settlement calculations has a decisive effect on the direct measurement of deformation parameters of the subsoil. It frequently contains a constitutive relation which is either verified or discharged by the direct measurement of deformation properties.

In principle, direct measurements can be divided into laboratory measurements on undisturbed samples and field measurements. Methods which directly call for laboratory measurements of undisturbed samples are: the stress path method, the state boundary surface method and the method of oedometric compression. Empirical methods of settlement calculations are, on the other hand, based on field tests. The third group of methods – the elastic method and the Skempton-Bjerrum method – allows the test methods or their combinations to be chosen at will; only more complicated cases (anisotropic and non-linear halfspaces) require laboratory tests.

In some cases laboratory and field tests are correlated – for example, the oedometric compressibility test and the static penetration strength test (Gielly et al., 1969; Sanglerat, 1972, p. 359). Then the method of deformation parameters' measurement can be chosen without regard to the method of settlement calculations. It generally holds that laboratory tests provide more information than field tests but their results are apt to be substantially distorted by sampling and handling of undisturbed samples.

(a) *Laboratory tests of undisturbed samples.* The results of laboratory tests depend on: the method of sampling (and of transport and storage); on handling and insertion of the sample in the apparatus; on the initial state of stress in the sample inserted in the apparatus; on parasitic effects of the apparatus; on systematic and random errors of measurement which affect the accuracy of the results; and finally, on the test method. These factors are mutually related and each depends on the nature, geostatic state of stress, etc. of the subsoil.

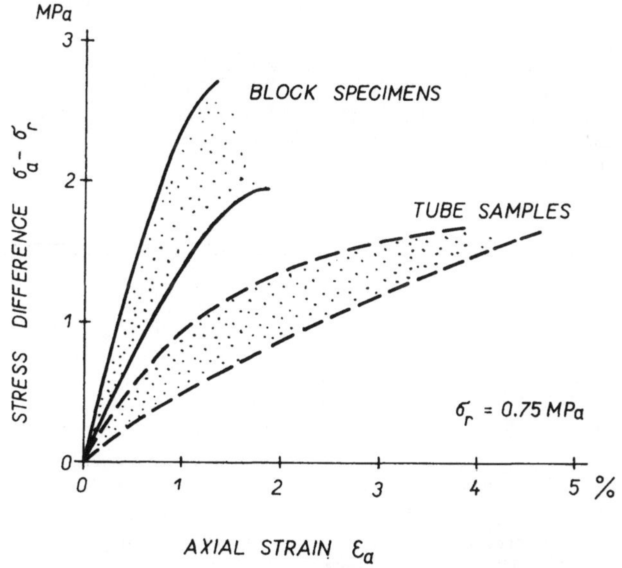

Fig. 54. Stress-strain diagrams of triaxial tests of London clay (vertical samples, site *P*) for hand-trimmed block specimens and tube samples (Ward et al., 1959).

(aa) Method of sampling: Fig. 54 shows an example of the effect of sampling techniques on the deformation modulus of stiff fissured clay. The deformation moduli of samples prepared from block specimens are on the average about four times those of tube samples. The difference is largest for very hard clays. (According to the writer's measurements, the deformation moduli of highly decomposed gneiss differed for such a case by only 1/3 – Feda, 1976c.)

The effect of sample disturbance during sampling is probably greatest for soft sensitive and stiff fissured clays. In the first case, small deformations are sufficient for disturbing weak, brittle diagenetic bonds; in the second, structure disturbance is made easier by fissures. Fissures are also instrumental in making the effect of sample size well-defined (Marsland, 1973); this effect is especially typical of jointed rocks, large samples of which are apt to have a deformation modulus many times smaller.

In the case of soft subsoils sample disturbance was reduced to an acceptable level by the introduction of thin-walled piston samplers, and by decreasing friction by means of foils*). When penetrated by fissures, stiff and hard subsoils require a considerable amount of cutting energy (the sampler must also cut through intact pieces of the subsoil). Along the fissures, on the other hand, the structure is greatly weakened. Under the circumstances it seems most suitable to take block samples (see also La Rochelle and Lefebvre, 1971).

The effect of subsoil disturbance by sampling frequently manifests itself quite strongly by a decrease of the deformation modulus E which is likely to be smaller by a factor of 3 to 13 than its probable in situ value**) (Burland, 1973). This effect is less strong on shear strength and possibly only slight on oedometric compressibility (Ward et al., 1959), except in the case of small pressures (<0.1 MPa – McGown et al., 1974). In oedometers, sampling disturbance is revealed by a shift of the compressibility curve towards smaller void ratios, and by a decrease in the pre-consolidation stress (Terzaghi and Peck, 1967, p. 78; McGown et al., 1974; Crawford and Eden, 1965, obtained pre-consolidation stresses in tube samples lower by 0.1 MPa than in block samples of Leda clay). In contrast to overconsolidated clays, the compressibility of normally consolidated clays is usually lower than the actual value as a result of sampling disturbance (Simons, 1974). This points to the possibility – and such cases have been noted – of an agreement between the laboratory and the field values of compressibility in some cases (Wilkinson et al., 1969). The effect of time is also important: the longer is the period between sampling and test, the more the sample is "disturbed" (Marsland, 1973; Bjerrum, 1973).

(ab) Preparation and placing of samples in an apparatus. Cutting a sample to apparatus' dimensions causes the structure in the plane of the cut to become

*) This holds particularly true in cases when samples from the middle section of the sampler cylinder are used in the test (La Rochelle, 1973).

**) This value is often estimated only from the measured magnitude of the settlement by back calculations.

more or less disturbed; in the contact with the rigid plate of a triaxial apparatus, and especially of an oedometer (in which the sample is thinner), the disturbed zone of the sample is exceedingly compressible (Zelst, 1948). This is the well-known bedding effect. In more favourable cases it amounts to about 5% of the axial deformation of the oedometric sample (Savvateiev, 1972). It is usually disclosed by the shape of the compressibility curve. Furthermore, the sample diameter can be slightly smaller than the diameter of the oedometer cylinder (Leussink, 1954), provided that the sample is not directly cut by the oedometric ring. When this is the case, the sample, before its structure collapses, is under unconfined compression. The same effect can also take place for other reasons, for example, in undisturbed samples of macroporous loess (Varga, 1965).

(ac) The initial state of stress of a sample in an apparatus. Its principal object is to reduce the effect of sample disturbance caused by sampling and preparation for the test and thus experimentally reconstruct its in situ compressibility curve. Re-consolidation to the original geostatic state of stress has been found suitable for high-quality undisturbed samples (Bjerrum, 1973, p. 120) including those taken from cemented clays (La Rochelle, 1973). In typical samples, on the other hand, this process leads to a decrease in their volume compared with that under in situ conditions. Hence the consolidation pressure must be reduced (for example, by carrying out isotropic consolidation by means of a cell pressure equal to 1/2 to 1/3 of the vertical geostatic pressure – Raymond et al., 1971), or alternatively, proceed without re-consolidation and correct the laboratory compressibility curve (Ladd, 1973; Schmertmann's correction – cf. Terzaghi and Peck, 1967, p. 78).

In overconsolidated clays the effect of sample disturbance can successfully be restricted by taking the reloading compressibility curve for the decisive one (Simons, 1963).

(ad) Parasitic effects of apparatuses, and other errors. In triaxial apparatuses, the term parasitic effects is applied to the effect of paper drains, rubber diaphragm and piston friction. In modern types of instruments, the last-named effect is eliminated by placing the load cell in the triaxial cell (between the sample and the piston); the other effects only play a role in more compressible samples and are taken care of by the application of an appropriate correction.

Whenever the surface of the samples is uneven (cohesionless soils), the measured volume deformation is distorted by the diaphragm being pushed into the peripheral pores of the sample (if $\sigma'_r \neq$ const). A suitable correction can also be introduced in this case.

In oedometers, the lateral friction of samples rubbing against their walls is an important parasitic effect which can, however, be reduced, by decreasing the roughness of the oedometer walls, to a value equal to less than 6% of the axial load (Leonards and Girault, 1961) and by lowering the sample height v. Usually $v/D \leqq 1/4$ is given as the permissible height-to-diameter ratio of samples, although $v/D \leqq 1/3$

(Savvateiev, 1972) or even $v/D = 0.4$ (Sanii, 1973) are also used sometimes. This discord can be explained by the fact that the magnitude of friction depends not only on the v/D ratio but also on the absolute height of the sample and its compressibility since the degree of friction mobilization is determined by the magnitude of the relative displacement between sample and oedometer wall (Feda, 1976b).

The effect of the apparatuses is manifested by making the oedometric compressibility depend on the size of the sample (Sanii, 1973; Feda, 1976b): for large sample dimensions, the scale effect is induced by skin friction, for samples of small thickness, by the bedding effect and sample non-homogeneity (Feda, 1976b). When the parasitic effects of the apparatuses are properly corrected, oedometric measurements yield practically the same compressibility as triaxial measurements – Fig. 55.

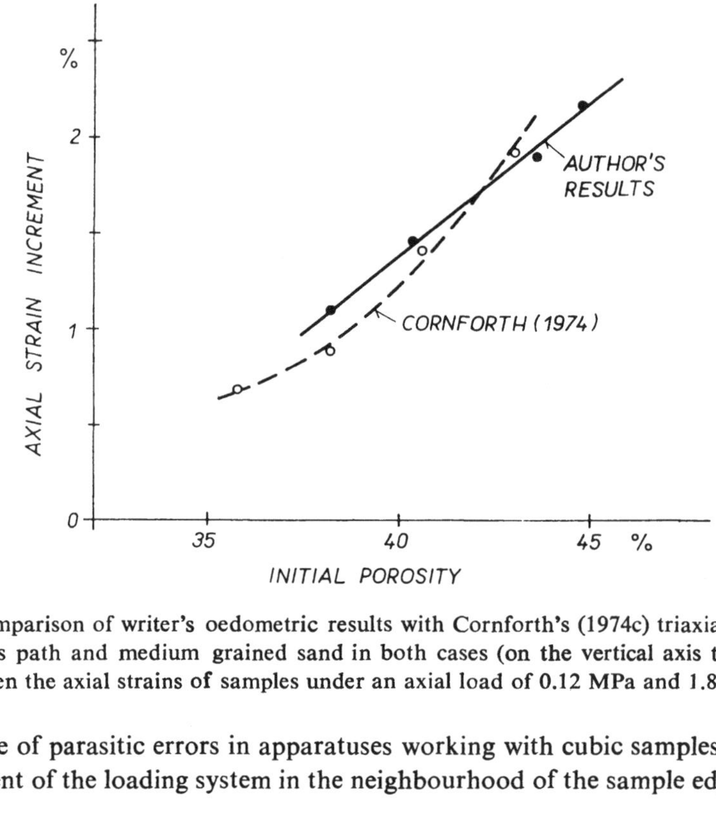

Fig. 55. Comparison of writer's oedometric results with Cornforth's (1974c) triaxial test results, for K_0 stress path and medium grained sand in both cases (on the vertical axis the difference between the axial strains of samples under an axial load of 0.12 MPa and 1.89 MPa).

A source of parasitic errors in apparatuses working with cubic samples can be the arrangement of the loading system in the neighbourhood of the sample edges (Green, 1971).

Next to systematic errors caused by imperfect apparatuses, techniques of measurement, etc., random errors such as subjective errors of the operators, fluctuations of temperature and moisture in the laboratory, can also play a part in laboratory measurements of deformation parameters.

(ae) Test methodology. This mainly refers to the system of loading or deformation of the sample, the loading (deformation) path and level.

As is well known, the magnitude of the load increment is apt to have a considerable effect on the shape of the compressibility curve (Leonards and Altschaeffl, 1964; Crawford, 1964); the mode of loading and of gradual destruction of the sample structure governs the reliability with which the preconsolidation load can be determined. Distinction is made between overconsolidation which is a result of the removal of overburden pressure, fluctuation of the groundwater surface, cementation, etc., and delayed consolidation (Bjerrum, 1967, 1973). As far as the settlement is concerned, the effect of overconsolidation is more favourable than the effect of delayed consolidation – the strength reserve which it imparts to the subsoil can only be used to about 50% in the second case (Simons, 1974).

When a soil deforms plastically, its structure changes and this in turn affects its deformation (and frequently also its strength) parameters. It is, therefore, generally necessary to assume that the path (anisotropy) and the level of stress and the rate of deformation determine (together with the original structure) the engineering behaviour of subsoils. The reverse can only be expected in soils with a comparatively strong structure (dense sands, cemented and overconsolidated clays, etc.) and/or at a relatively low stress level.

Fairly ample experimental material is available to prove the considerable effect of the stress path and level (if the stress paths are linear, the examination is concerned with the combined effect of stress anisotropy and level) on the deformation parameters E and μ (see, e.g., Breth et al., 1973), and possibly on the parameters a, n, R_f in eqns. (124) and (125) – Yudhbir and Varadarajan, 1975; Desai, 1975 – or on the shape of the stress-strain diagram (Lomise et al., 1969; Green, 1971; Campanella and Vaid, 1973), especially when comparing triaxial tests to plane strain tests of samples (see, e.g. Vaid and Campanella, 1974).

So far no apparatus has been developed which would permit laboratory simulation of a general stress path (including the rotation of the principal axes of the stress tensor), and even if such an apparatus were available, it would be of purely scientific significance. From an engineering point of view, it is more to the point to start from the results of standard tests and possibly, to use corrections, depending on the kind and conditions of the subsoil, whenever the deviations of the stress path and level from the standard equivalents are fairly substantial.

Several partial conclusions can be drawn on the basis of this rather informative review of problems associated with laboratory measurements of mechanical parameters of subsoils: the test results must always be critically appraised, interpreted and, if required, corrected to take into account the kind and conditions of the subsoil and of the structure for which the settlement is calculated; notwithstanding comparatively great attention devoted to laboratory tests, witnessed by the number of relevant publications, questions rather than exact answers are known at present; in the case of certain foundation soils, one must give up the possibility of performing laboratory tests, probably because of their structure and/or texture (large disturbance of samples, impossibility of sampling, the scale effect, etc.).

(b) *Field tests.* Besides the field tests which were discussed in Section 3.7 and which will not be dealt with here, the plate loading test is the fundamental field test.

It is usually assumed that the subsoil immediately beneath the plate is homogeneous, and E is calculated using eqn. (89) (for a chosen value of μ); however, the assumption of non-homogeneity can also be applied (Carrier and Christian, 1973). Since plate loading tests can be made at different depths, they can be used to determine the variations of the deformation modulus with depth.

The test results are affected (Marsland, 1973) by: the degree of subsoil disturbance due to the effect of boring (it is recommended to excavate and clean by hand the surface of the subsoil prior to the embedment of the plate); the time between the denudation of the subsoil and the test; structural changes of the subsoil produced by the effect of its unloading (for example, opening of fissures); the size of the plate (the plate must be large enough so that the subsoil disturbance immediately below the base of foundation of the plate will play no important role – for example, 86.5 cm for London clay and chalk – Marsland, 1973; Burland et al., 1973); the size of the borehole and its lining which decides about the correction factor (Fig. 32).

The shortcomings of the plate loading tests are as follows: a certain mathematical model (most frequently, a quasi-linear, isotropic and homogeneous) of the subsoil is a priori assumed in the evaluation; the magnitude of μ is assumed rather than measured; and usually only a single deformation parameter, E, is determined. More complicated field tests, for example, triaxial, are rarely carried out. Hence, the mechanical parameters of only the simplest halfspaces – isotropic, homogeneous and non-homogeneous – can as a rule be defined by plate loading, and the test itself is of a model character.

As this review of laboratory and field tests of deformation parameters of subsoils indicates, both kinds of tests have advantages as well as disadvantages so that they supplement rather than replace one another. It is, therefore, interesting and instructive to compare some of their results.

The deformation modulus determined by means of a loading plate may agree with the pressiometric deformation modulus (Meigh and Greenland, 1965; Ménard, 1971; Hobbs and Dixon, 1969).

The deformation moduli of rocks measured on small samples or by means of small-diameter loading plates, are substantially (ten times and over) larger than those of rock masses. Here the scale effect plays a part because a rock mass, in contrast to the sample, is penetrated by a greater number of surfaces of weakness which increase its deformation.

With decreasing total strength of the subsoil, this difference gradually grows smaller. In chalk sediments Burland and Lord (1969) obtained a ratio between the laboratory deformation modulus ($D = 5.5$ cm, sample cut from a block, small effect of σ_r') and the field deformation modulus, equal to 2 (the laboratory $E \sim 6300$ MPa, $\mu = 0.24$). Similarly, a ratio of about 3 (samples with $D = 2.5$ cm) was obtained by Meigh and Greenland (1965) for Bunter sandstone.

In more deformable subsoils in which the degree of sample disturbance comes into play gradually, the differences between E_{lab} and E_{field} (i.e. E_{plate}) are not great, provided that the sample disturbance is not excessive or that it is successfully eliminated by means of sample re-consolidation.

This is clearly seen in Fig. 56, showing the case of a stiff fissured clay of glacio-lacustrine origin. The deformation moduli of block samples which are least disturbed

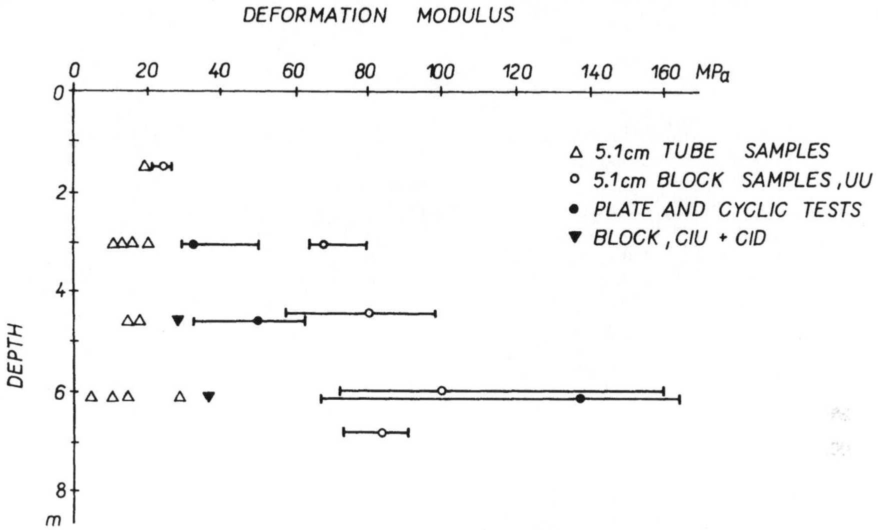

Fig. 56. Comparison of deformation moduli of undisturbed samples (block specimens and tube samples) and of soil subjected to field loading test — stiff fissured Ontario clay (Lo et al., 1971). Symbols used for test designation: UU — unconsolidated undrained tests; CIU — isotropically consolidated undrained test; CID — isotropically consolidated drained test.

for the given subsoil when judged according to the above criteria, are in good agreement with the deformation moduli established by means a of loading plate ($D = 0.3$ m).*) This fact is especially borne out by cyclic-load laboratory tests for which the specimens were also prepared from block samples. An agreement between the laboratory and the field deformation moduli is also reported by Meigh and Greenland (1965) for tests of Bunter sandstone, and by Dalmatov et al. (1973) – see Fig. 63.

Because of sampling disturbance, tube samples (Fig. 56) yield values of E which are smaller by a factor of 4–7. This agrees, as to the order of magnitude, with the result shown in Fig. 54. Since the subsoil is gradually becoming stronger, the differences increase with depth. This is demonstrated by the results of Cornforth (1974a) obtained for Keuper Marl – starting at 34%, the divergence gradually increases with decreasing moisture content.

*) Since the measured E_h/E_v ratio was equal to 1.4 in this case, it would have been necessary to consider the subsoil anisotropy in the evaluation of the loading tests.

At a depth of 1 – 2 m at which the subsoil is comparatively least strong, the difference between the deformation moduli (Fig. 56) is negligibly small (about 25%). Besides the quantitative differences in the magnitude of E, one also finds that E varies differently with depth. While block samples and the results of loading tests indicate subsoil non-homogeneity, tube samples yield E = const. with depth. This is also claimed by Cornforth (1974a) for Keuper Marl.

A similar conclusion can be drawn on the basis of Fig. 57. The deformation moduli determined by loading tests were 1.8 to 4.8 times larger than the laboratory ones. The

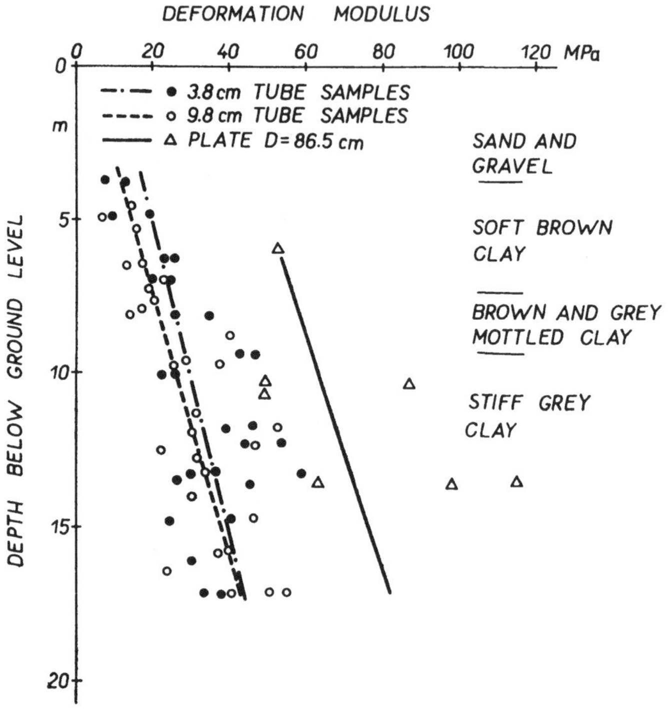

Fig. 57. Comparison of deformation moduli obtained by laboratory and field tests of stiff fissured London clay (Marsland, 1973).

deformation moduli of small-diameter (3.8 cm) samples are up to 1.3 times larger than those of large-diameter (9.8 cm) samples; this is a clear proof of the scale effect which can be expected in fissured media.

A disagreement between the laboratory and the field deformation moduli can also be noticed in other than clayey subsoils, provided that they are sensitive to disturbance. Thus, for example, Varga (1965) found the deformation modulus of subsident loess obtained by means of a small-plate field test to be about seven times the laboratory value which was about 1/10 of the deformation modulus ascertained by means of a field test made with a large plate of 0.5 m^2 area (in the latter case, E = 55 MPa).

The extensive material collected in the course of standard field plate loading tests and oedometric laboratory tests was analyzed by Agishev (1957), who processed a total of 87 loading tests (made with plates having an area from 2500 to 10 000 cm^2 in trial pits, and an area of 600 cm^2 in boreholes drilled to 12 m depth) of plastic Quarternary clayey soils, and 437 oedometric tests of genetically different plastic clayey soils. Although the comparison has a number of shortcomings (genetically different soils, no distinction made between the effects of depth, loading plate size, etc.), the correlations obtained were used for calculating the deformation modulus (for $\mu = 0.3$, $K_0 = 0.65$, assuming normal consolidation). As the results shown in Fig. 58 imply, the ratio of the field deformation modulus to the laboratory value is

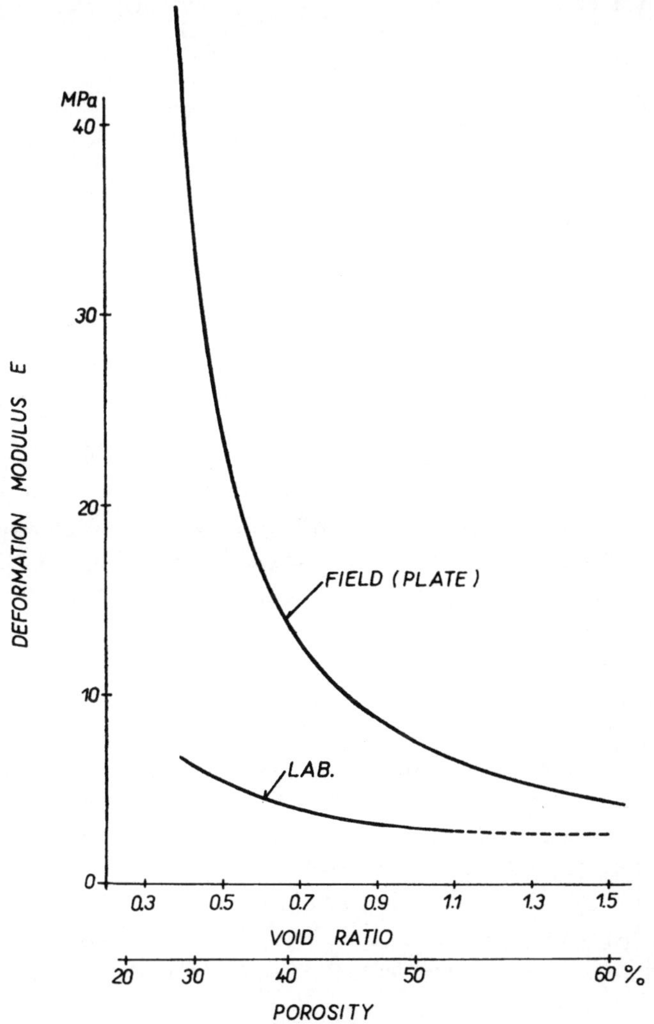

Fig. 58. Comparison of deformation moduli obtained by laboratory and field tests of plastic clayey soils (after Agishev, 1957).

2.4–4.6 and increases with the strength (decreasing void ratio) of the subsoil.*) This conclusion generally agrees with the above statement as well as with Fig. 56 (the void ratio decreases with depth).

Fig. 58 makes it possible to obtain the magnitude of the correction factor (>1) by which one must multiply the oedometric deformation modulus in order to establish a deformation modulus typical of the corresponding field test. According to Shvets' (1964) experiments, the value of this correction factor is the same for all clayey sediments, and according to Gorbunov-Posadov and Sinelschikov, it varies from 10 ($e = 0.4$) to 2 ($e = 1.6$).

The revised values of this correction factor as recommended by Ignatova (1968) are set out in Table 2.

Table 2

Type of soil	Void ratio										
	0.41 to 0.5	0.51 to 0.6	0.61 to 0.7	0.71 to 0.8	0.81 to 0.9	0.91 to 1	1.01 to 1.1	1.11 to 1.2	1.21 to 1.3	1.31 to 1.4	1.41 to 1.5
Sandy loam	4.2	4.2	3.7	3	2.2	—	—	—	—	—	—
Loam	5	4.8	4.5	3.9	3.2	2.6	2.1	—	—	—	—
Clay	—	—	6	6	5.8	5.4	4.8	4.1	3.4	2.7	2

As Fig. 58 suggests, at a high enough porosity (approximately 70%) the two moduli are identical. This limit approximately corresponds to soft clays and is in agreement with the previous statement according to which the greatest disturbance is observed in samples taken from stiff to hard fissured clays. It goes without saying that it also depends on the quality of sampling. If it is assumed, in keeping with Fig. 56, that block samples yield the same E as field loading tests, then the results published by Cornforth (1974a) for Keuper Marl indicate that the effect of tube sampling disturbance plays a role only when the moisture content is lower than the plastic limit (about 34% in the case being considered), i.e. in samples of subsoils in stiff state.

In conclusion, it seems to the point to mention the accuracy with which a deformation modulus can be measured. Errors associated with the measurement proper do not exceed 5% (Kulchickiy, 1969). On the basis of a laboratory analysis and in particular, field plate loading test data of a supposedly homogeneous subsoil, Konovalov and Rudnitskiy (1964) give the ratio of the maximum to the minimum deformation modulus as 1.4 (sand), 1.8 (Quarternary clays, loams, sandy loams) and 2.5 (clayey moraine sediments). A variability coefficient of 16%–43% (which clearly also includes

*) In cases when the void ratio is less than 0.5, one can hardly speak of a plastic state of the tested soils, as Agishev (1957) does.

a certain subsoil non-homogeneity) corresponds to these values. By its order of magnitude, this coefficient agrees with the variability coefficient of 20%–40% reported by Schultze (1975). For plastic Quarternary clays, the data of Agishev (1957) lead to a variability coefficient of the deformation modulus of about 25% for field plate loading tests and a subsoil porosity of about 50%. About a half of this value is claimed by Naborczyk and Pałka (1975) for silts and silty clays. It is worthy of mention that an inaccuracy in the determination of E up to 30% has virtually no effect on the distribution of the contact stresses, bending moments, shear forces and deflections of a foundation (Gorbunov-Posadov and Malikova 1973).

A useful review of measured deformation parameters of various types of subsoils was published by Gerrard (1975). As already mentioned, tabulated values of the deformation parameters E and μ can be found in many Standards (for example, the Czechoslovak Standard ČSN 73 1001, the Polish Standard PN-59-B-03020, the Soviet Standard SNiP II-B.1-62 and SNiP II-15-74, Part II, Chapter 15).

Table 3

Soil	Φ_f low σ_r	Φ_f high σ_r	c MPa	a	n	R_f	g	f	d
Gravel well graded (GW)	47	35	—	500	0.3	0.7	0.32	0.14	6.4
Gravel poorly graded (GP)	46	38	—	1800	0.3	0.8	0.38	0.15	8.0
Sand well graded (SW)	50	35	—	300	0.5	0.7	0.3	0.13	5.0
Sand poorly graded (SP)	40	30	—						
Oroville dam:									
Amphibolite gravel	43.5		—	3780	0.19	0.76	0.43	0.19	14.8
Amphibolite gravel finer	43.5		—	3350	0.19	0.76	0.43	0.19	14.8
Clayey sandy gravel (GC)	23.1		0.126	345	0.76	0.88	0.30	−0.05	3.83
Soft clay	13.0		0.029	150	1.0	0.9	0.49	0	0
(Concrete)	0		20.7	137500	0	1.0	0.15	0	0
Thomastown dam:									
Clayey sand (SC)	17.0		0.086	30	0.94	0.61			
Canyon dam:									
Low to medium plasticity clay (CL)	3.0		0.2	175	0.41	0.90			
Pittsburgh sandy clay (SC)	14.0		0.103	40	0.48	0.68	0.43	−0.05	0.6

Note: Cohesionless soils were tested under drained conditions, cohesive soils under unconsolidated, undrained conditions.

For hyperbolic substitution of the stress-strain diagrams according to eqns. (124), (125), (126) and (127), Table 3 (after Kulhawy and Duncan, 1972, and quotation by Desai and Abel, 1972, p. 322) shows informative values of the corresponding parameters.

3.10 Analysis of some results of settlement measurements

Many comparisons of measured and calculated values of settlement have been published, especially in proceedings of international conferences exclusively devoted to problems of settlement (for example, London 1955; Wiesbaden 1963; Cambridge 1974). It can be stated that each method of settlement calculation can be documented by examples of its successful application. Thus, for example, the applicability of the elastic theory to the case of a homogeneous halfspace is proved by means of Polshin's loading tests by Gorbunov-Posadov and Malikova (1973, p. 16), to the case of a non-homogeneous halfspace for a number of structures and heaving of a foundation pit bottom by Butler (1974), to an anisotropic (Westergaard's) halfspace by Stefanoff et al. (1965). The stress path method was verified by Lambe (1963), the method of oedometric compression by Burland (1969b) (cf. Fig. 47), Lambe (1973), Huder and Sevaldson (1963) and others, the method of Skempton and Bjerrum by both its authors (1957) as well as by Butler (1974), the empirical methods by Sherif (1973), Schultze and Sherif (1973), Schmertmann (1970), Ménard (1971), Bru et al. (1973), etc. The finite element method was successfully applied by Burland et al. (1973), Dies et al. (1973), and others – see also Section 3.8. Statistical comparisons of the ratio between measured and calculated settlements which have also been published (for example, by Polshin and Tokar, 1957; Varga, 1965) have, on the other hand, shown the possibility of errors of several orders of magnitude.

It is clearly seen from this brief summary that the accuracy of settlement prediction can hardly a priori be related to the accuracy of the applied method. Using several examples, particularly the case of the Northeast test embankment, Lambe (1973) convincingly showed the simplest methods to be capable of yielding the most accurate predictions of both the initial and the final settlements while the finite element method, provided that it is not supported by actual experience, is likely to lead to very inaccurate results (oil storage dam – horizontal deformation).

If only a measured settlement of an object is available, it is valueless for testing any method of settlement calculations for the following three reasons:

– An agreement between a single measured and a calculated value can usually be achieved by a suitable manipulation of nearly any method – the problem has too many free parameters which are apt to appeal to the fantasy of those making the comparison. Published comparisons are in fact mostly predictions of the type C1 (Lambe, 1973) – the results are known in advance (in the opposite case, values even more than 10x in excess of the measured settlement are reported – Wroth, 1977).

– The deformation parameters which are the actual basis of a calculation, are seldom chosen with reliability. Field and laboratory tests are not carried out simultaneously, the effect of sample disturbance, size, etc. is not critically evaluated, and, what is worse, the scatter of the test results is frequently omitted. Although the choice of the deformation parameters is usually the most important factor of all, it is more often than not pushed in the background by the method of calculation itself.

– A possible agreement between a measured and a calculated settlement does not yet prove the correctness of the mathematical model on which the calculation is based. One can cite cases of such an agreement when the mathematical model was obviously incorrect, but in the cases under consideration the errors cancelled each other out (Lambe, 1973; cf. also Section 2.2.1).

A suitable method of settlement calculations is a method based on a mathematical model whose principal features comply with the physical reality of the foundation and the subsoil. In order to verify the existence of such conformity, one should

– measure the settlement not only of the foundation but also of the subsoil at various depths,

– measure the deformation of the surface, and possibly of other planes in the subsoil, produced by the settlement,

– determine, accurately enough, the in-situ deformation properties of the subsoil at various points.

The published material permitting, the above points of view will be employed in the present evaluation of the various methods of settlement calculations.

Numerous methods are available for measuring the displacements (settlement) at various points of the subsoil (cf., for example, Perlea, 1969). First measurements of this kind under model conditions were probably performed by Aichhorn and Gerber about 1930 (cf. Section 2.8.2.2) and since then by a number of researchers (for a partial review see Kriegel and Wiesner, 1973, and Naborczyk and Pałka, 1975). First tests of this kind relating to the subsoil of structures (the case of a 150 m high chimney stack) were described (and subsequently supplemented) by Egorov (1959; Yegorov and Nichiporovich, 1961; Egorov and Popova, 1971a). Similar measurements of both spread footings and pile foundations (for the latter case see Bartolomey, 1975) have since then appeared in the literature.

A selection of such measurements made on actual structures (with the sole exception of those of Tugaenko and Khutoryanskiy, 1971, who performed in situ loading tests using a 3 × 3 m plate) of various widths ($B = 13.5$ to 23 m, diameter of circular foundations $D = 9.3$ to 60 m) and depths of foundation ($d_z = 1.7$ to 13 m) on cohesive, cohensionless and layered subsoils, is shown in Fig. 59 (for further examples see Feda et al., 1977). The figure also shows the variation of the relative settlement with depth (w_0 – the settlement of the base of foundation, w_z – the settlement at depth z below the base of foundation) for a homogeneous elastic subsoil with

$\mu = 0$ to 0.5 and a uniformly loaded circular area, reported by Ahlvin and Ulery (1962) (the case of $\mu = 0$ corresponds to uniaxial compresion for $K_0 = 0$ – cf. the third of eqns. (13) and eqn. (42)). The uniaxial compression was generally (i.e. $K_0 \neq 0$) calculated using eqns. (91), (42), (106) and numerical integration. The difference between the results thus obtained and those corresponding to the curve $\mu = 0$ was slight – see Fig. 59.

It is seen in Fig. 59 that the variations of w_z/w_0 with depth in no case correspond to the elasticity theory or to the uniaxial compressions (for a linear isotropic and

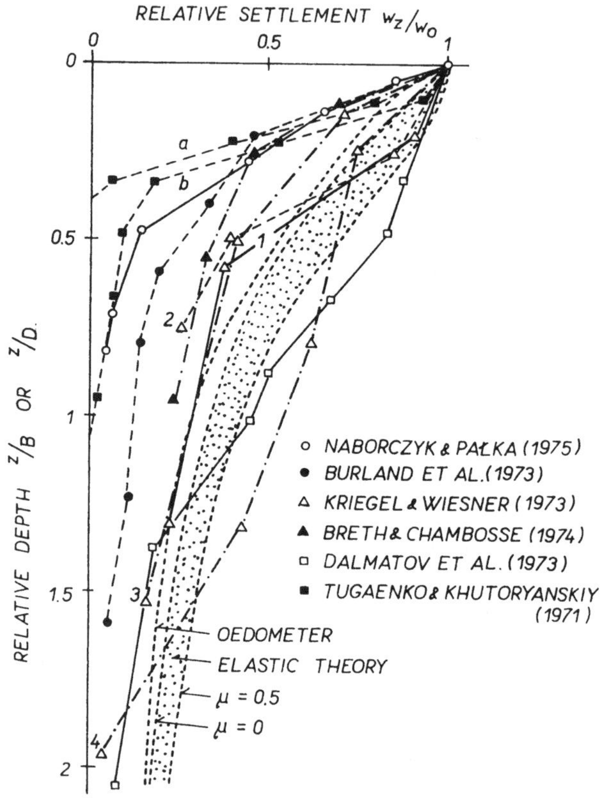

Fig. 59. Settlement at various depths beneath foundations (when several loads were applied, only values corresponding to the maximum load are presented) — measurements by: Burland et al. (1973 – diameter $D = 18.3$ m, foundation depth $d_z = 1.7$ m, chalk); Naborczyk and Pałka (1975 – $B = 14.6$ m, $L/B = 2.53$. $d_z = 2$ m, silts and silty clays); Kriegel and Wiener (1973 – 1: width $B = 23$ m, $L/B = 2.09$, $d_z = 13$ m, sand; 2: $B = 19.4$ m, $L/B = 1.52$, $d_z = 1.9$ m, sand; 3: $B = 20$ m, $L/B = 1.28$, $d_z = 3.3$ m, boulder clay; 4: $D = 9.3$ m, $d_z = 2.7$ m, boulder clay); Breth and Chambosse (1974 – $D = 60$ m, $d_z = 5$ m, sand and gravel); Dalmatov et al. (1973 – $B = 13.5$ m, $d_z = 2$ m, sandy and clayey soils with organic admixtures); Tugaenko and Khutorianskiy (1971 – $B = 3$ m, $L/B = 1$, $d_z = 1$ m, fill, clay, sand; a – $q = 0.08$ MPa, b – $q = 0.2$ MPa).

homogeneous halfspace). The most obvious difference is the rapid decrease of settlement with depth – at a depth z/B $(z/D) > 2$, the settlement is negligible in all cases. This depth to which the subsoil practically deforms is called the active depth (it was already mentioned in Sections 3.1, 3.2 and 3.5).

Model and loading tests performed with smaller loading areas suggest the conclusion that the depth of the active zone depends on the magnitude of the load (cf. the curves *a* and *b* in Fig. 59; for a circular foundation of 5000 cm^2 area on sand, Konovalov, 1971, finds direct proportionality between the depth of the active zone and the load q; the depth of the active zone increases with decreasing deformation modulus and decreases with decreasing foundation area). Beneath actual structures, on the other hand, the depth of the active zone is virtually insensitive to the magnitude of the load (in the usual range) – Kriegel and Wiesner (1973), Naborczyk and Pałka (1975) – and it is found to be invariable in time (Konovalov, 1971).

This disagreement between model tests and prototypes as regards the sensitivity of the depth of the active zone to the magnitude of the load leads to a rather sceptical opinion of the significance of model tests (Kriegel and Wiesner, 1973; Naborczyk and Pałka, 1975). The simplest explanation of the rapid decrease of vertical displacements with depth is the fact that the deformation modulus of the subsoil increases with depth. Then two foundations of different dimensions are similar only if G_0/aB defined by eqn. (69) is the same in both cases. This stipulation requires that the increase of the deformation modulus with depth should be smaller by the same factor by which the width or the diameter of the foundation is larger (Fig. 64). If the subsoil is the same, then it is less compressible and hence less sensitive to small changes in the stress σ_z and to the effect of time for larger foundations. Analogous arguments can be put forward in the case of a non-linear halfspace (while a linear, a homogeneous and an anisotropic halfspace satisfy model similarity). The cause of subsoil non-linearity is similar to the cause of subsoil non-homogeneity – an increase of the geostatic stress and a decrease of the induced stress increment with depth; thus the two mathematical models express the same physical reality – an increase of the deformation modulus with depth – but since it is substantially simpler, the first, a non-homogeneous halfspace, is preferred (cf. Sections 2.7 and 2.9).

Hence the principal conclusion which can be drawn on the basis of Fig. 59 is that the only realistic method of settlement calculations is a method which makes it possible to take into account the dominant physical feature of subsoils – their non-homogeneity (natural or stress-induced). The difference between the uniaxial and the triaxial mechanisms of compression plays an irrelevant role in this respect.

The irregularity of some of the settlement curves shown in Fig. 59 is a result of the effect of variously compressible layers of the subsoil. Thus, for example, the slope of the curve *b* increases in the range $z/B = 0.1$ to 0.33 in which the subsoil is strongly compressible – a layer of organic clay. Similarly, the settlement curve after Dalmatov et al. (1973) has two regions of a sudden increase of the slope, i.e. for $D/B = 0.5$ to 0.9 and for $D/B = 1$ to 1.4. The first corresponds to a layer of organic sandy soil, the second to soft varved clay. In the case of curves obtained by Burland et al. (1973), Naborczyk and Pałka (1975) and Breth and Chambosse (1974), whose curvature changes gradually, one can, on the other hand, suppose that the subsoil is continuously non-homogeneous. The agreement between some of the settlement curves and the elasticity theory (the range of curves with $\mu = 0$ to 0.5) can be explained by a less compressible subsoil immediately beneath the base of foundation.

Referring to Fig. 59 it is seen that the problems of settlement calculations concentrate on a clear description of continuous or discontinuous non-homogeneity of the subsoil and, naturally, on a reliable determination of the deformation parameters of this non-homogeneity. A simple engineering method of introducing this non-homogeneity in the calculations is to assume that the depth of the active zone represents the thickness of the compressible foundation soil. In this way a discontinuously non-homogeneous halfspace is formed as a model of a seemingly homogeneous halfspace. The stress σ_z is regarded either as unchanged (cf. Section 2.5.2) or as modified by the presence of a compressible layer of finite thickness (Egorov, 1959; Yegorov and Nichiporovich, 1961; Egorov and Popova, 1971a,b, etc.).

The problem that remains is the determination of the thickness of the active zone, d_{act}. The relative depth of the active zone depends on subsoil non-homogeneity*), and the only reliable way is to deduce it from the calculation of w_z, which takes this non-homogeneity into account (Fig. 37; Feda et al., 1977). The many simple methods which are available are approximate or roughly approximate:

– The term "active depth" is taken to denote a depth at which the increment of σ_z produced by the structure is less than 20% of the vertical geostatic stress. It is found that this limit actually ranges from 25% to 90% and even to 100% (Shvets and Kulchickiy, 1971).

– A constant thickness of the active zone is recommended (see a summary by Sherif, 1973, p. 43 et seq.), very frequently a value $z/B = 2$ (Sherif, 1973, p. 43; Schmertmann, 1970; Czechoslovak Standard ČSN 73 1001) which represents the upper limit according to Fig. 59. The depth of the active zone for circular ($D > 20$ m) foundations of chimney stacks recommended by Egorov and Popova (1971a,b) is $D/2$ for clayey subsoil and $D/3$ for sandy subsoil (if a strongly compressible layer exists at a depth $> D/3$, the active zone must be enlarged by its thickness). It is clearly seen in Fig. 59 that $z/B = 2$ is often a conservative value; the settlement curve after Breth and Chambosse (1974 – sand with gravel and $D = 60$ m) shows the active zone to be substantially larger than that of Egorov and Popova (1971a,b). Hence the recommendations of the latter authors are apt to lead to underestimates of the settlement.

*) As laboratory tests made with sand have shown, d_{act} depends not only on density but to a significant degree also on the way in which sand has attained this density — Dovnarovich, 1975. Hence porosity is not the only structural parameter which matters (see also Ladd et al., 1977, p. 433).

— Malikova (1972) determines the thickness of the active zone by calculating the settlement of a homogeneous layer of finite thickness and by requiring that this settlement should be equal to the measured settlement (this author examined a set of 20 objects). For fairly wide foundations ($B \geqq 10$ m) with $L/B = 1-6.5$, $q = 0.1-0.65$ MPa and $E = 15$ to 60 MPa (plate loading tests at depth), Malikova obtained the following empirical relation for the depth of the active zone d_{act}:

$$d_{act} = \sqrt[4]{\frac{B}{\alpha}} \tag{129}$$

where the parameter $\alpha = 6 \times 10^{-4}$ m^{-3} for clayey subsoils, and 3×10^{-3} m^{-3} for sandy subsoils, B — the foundation width in m. According to eqn. (129), d_{act} decreases with foundation width (d_{act}/B is approximately in indirect proportion to B), just as established by Konovalov's (1971) tests. Because of its empirical nature, eq. (129) is, however, of only regional validity.

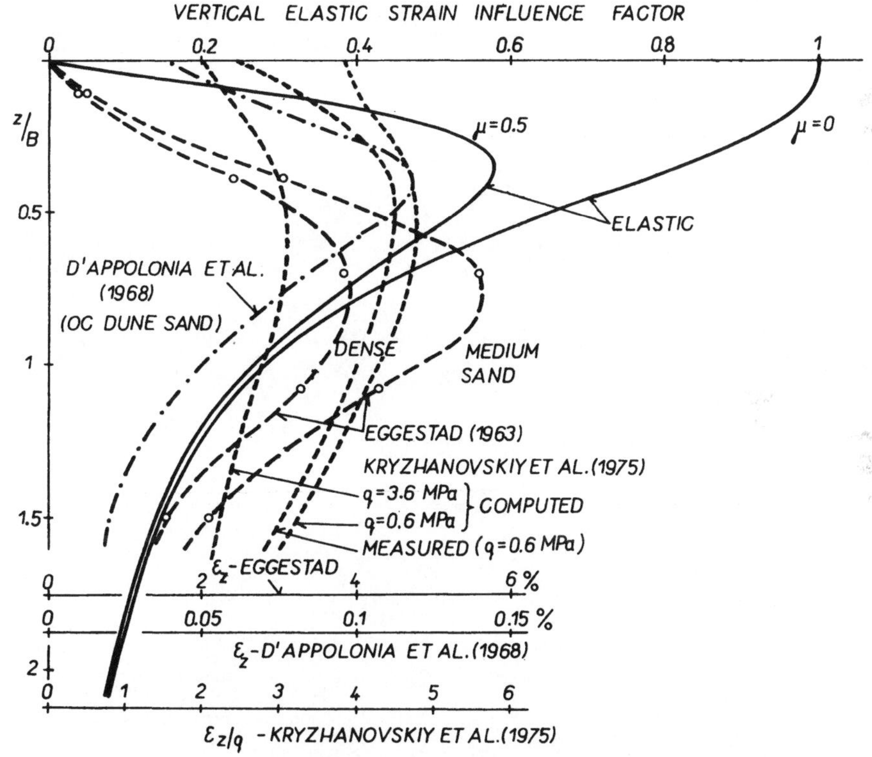

Fig. 60. Relation between vertical strain ε_z and depth according to elasticity theory, model tests of Eggestad (1963) and Kryzhanovskiy et al. (1975 — cf. Fig. 49) with sand, triaxial stress path tests on overconsolidated dune sand — D'Appolonia et al. (1968), and according to calculations by the method of finite differences (Kryzhanovskiy et al., 1975).

– Altes (1976) statistically analysed about 140 evaluations of d_{act} and deduced a formula relating d_{act} to the foundation width, length and area, normal contact pressure and the deformation modulus of the subsoil. Since the majority of cases treated by him were either laboratory and plate loading tests or the value of d_{act} was deduced indirectly, like eqn. (129), his relation has to be verifield for actual foundations by direct measurements.

Fig. 59 corresponds to Fig. 60 which shows the vertical deformation ε_z as it depends on depth (its integration yields the curves of w_z shown in Fig. 59). A comparison with numerous model tests made with sand shows the experimental curve to be qualitatively close to the theoretical curve (for a homogeneous subsoil) with $\mu > 0$. Referring to Fig. 59 (the curves 1, 3a,b) one can deduce from this fact that in homogene-

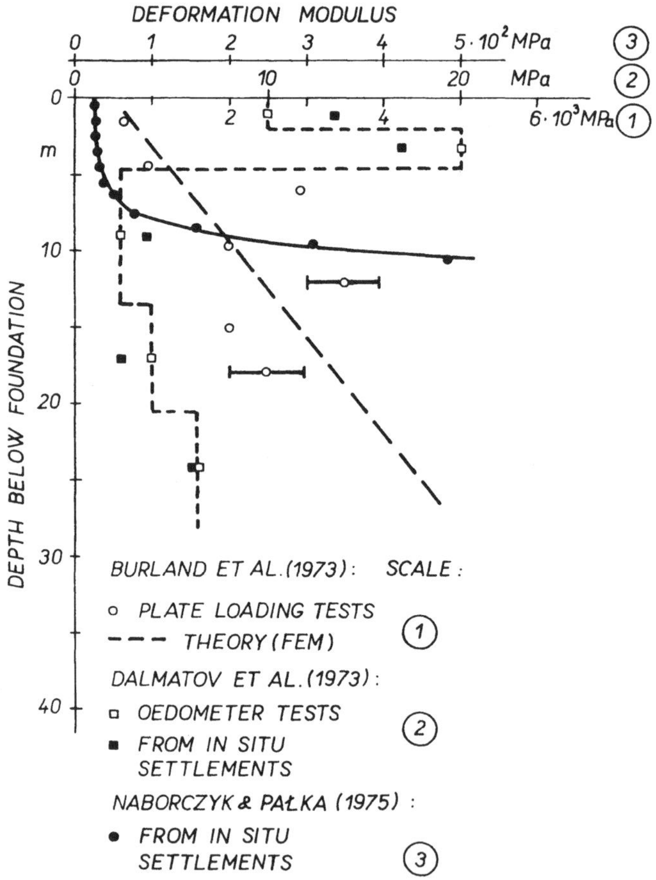

Fig. 61. Variations of the deformation modulus of subsoil determined by means of a plate and the finite element method (Burland et al., 1973), oedometric tests (Dalmatov et al., 1973), and calculated from in-situ measurements of settlement of individual layers cf subsoil (for Boussinesq's state of stress) — Dalmatov et al. (1973), Naborczyk and Pałka (1975).

ous sandy subsoils, lateral spreading is apt to contribute significantly to the total settlement as a result of a comparatively high Poisson's ratio. In cohesive foundation soils in which μ is likely to be lower (Wroth, 1971 – $\mu' = 0.12$ for London clay), the contribution of compression is usually more important (in immediate settlement, μ is, of course, 0.5). But even to sandy subsoils, the results of model measurements shown in Fig. 60 can hardly be applied without reservations. This is borne out by measurements of Breth and Chambosse (1974) presented in Fig. 59, as well as by those of Dunn (1974)*).

If a conclusion concerning the deformation modulus of the subsoil generally increasing with depth was drawn on the basis of Fig. 59, it may do well to subject it to further verification. The measurements and their analysis published by Burland et al. (1973) (short-term settlement of a tank on a chalky subsoil) are probably the most complete of all in this respect. Fig. 59 is complemented by Figs. 61 and 62. The measured settlement shown in Fig. 59 was checked numerically by means of an analytic method

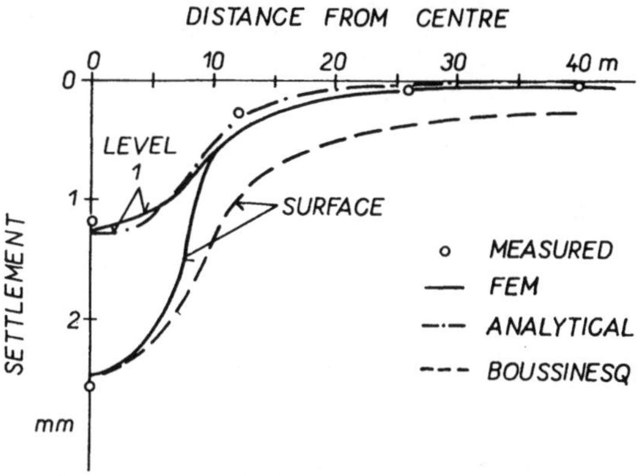

Fig. 62. Calculated (by means of the finite element method and analytically according to Awojobi and Gibson, 1973) and measured settlement profiles after Burland et al. (1973) – cf. Figs. 59 and 61.

(according to Awojobi and Gibson, 1973, assuming complete non-homogeneity as defined by eqn. (73)), and by means of the finite element method. The variations of the deformation modulus obtained by the latter method are shown in Fig. 61 in dash lines. They were proved to be correct by plate ($D = 86.5$ cm) loading tests. As indicated by Fig. 62, unlike in Boussinesq's solution, the deformations of the surface and of the level 1 also corresponded very well to the numerical values**). The mathematical model of a continuously non-homogeneous subsoil (E increasing linearly with depth) is thus perfectly verified.

*) As demonstrated by loading tests – Marsland and Eason, 1973; Marsland, 1973 – the subsoil immediately beneath the base of foundation is frequently disturbed and even remolded and, therefore, more compressible.

**) A similar difference between the actual deformation of the surface and the deformation calculated according to Boussinesq can also be noted in the case of pile foundations (Hlaváček, 1975) and during model tests – Shvetsov (1967). As shown by Dovnarovich's (1975) tests, the surface heaves when the subsoil is of small thickness.

If one measures the compression of the individual layers of the subsoil, one can calculate from it (under the assumption accepted in Section 2.5.2 that the stress does not differ too much from the stress in a homogeneous halfspace) the deformation moduli of these layers. Two such calculations are shown in Fig. 61. In the case of measurements made by Naborczyk and Pałka (1975) when the subsoil was continuously non-homogeneous according to Fig. 59, hyperbolic non-homogeneity defined by eqn. (70) (where $d = d_{act}$) was obtained. Hence an assumption of this kind can even be more realistic than the assumption of linearity according to eqn. (69).

In the second case – measurements of Dalmatov et al. (1973) – one notices discontinuous non-homogeneity: beneath a layer of less compressible sand several metres thick, the deformation modulus increases linearly with depth. A comparison with oedometric tests shown in Fig. 63 indicates a very good agreement.

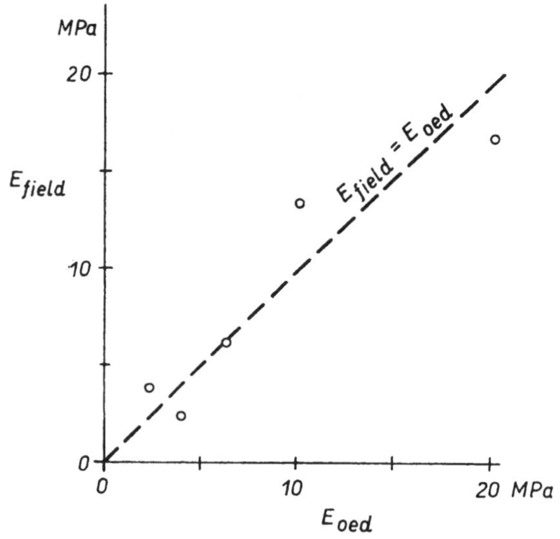

Fig. 63. Comparison between oedometric and field deformation modulus derived from measurements of settlement of gauge marks at various depths of subsoil (data by Dalmatov et al., 1973).

3.11 Model tests

The problem of modelling of structures from the point of view of their settlement deserves a brief notice. If one succeeds in constructing a model of the structure and its subsoil, the settlement of the structure can be determined from the measured settlement of the model. Thus this method can also be classified among the methods used to determine the final settlement of structures, and the first empirical method described in Section 3.7 is its adaptation of a sort.

In order that the results of model measurements might be applied to a structure (its prototype), the model must satisfy a number of conditions.

A model of a foundation must be large enough relative to the dimensions of the elementary particles and the macrotexture of the subsoil (for example, the system of fissures, etc.). From this point of view, the minimum dimension of a model foundation on sand should be, for example, several centimetres.

In contrast to the in situ conditions, the volume of the model medium is much more restricted. Its boundaries (the bottom and the sides of the test box) must be at a distance so large that they exercise practically no effect on the deformation of the subsoil, i.e. the thickness of the subsoil must considerably exceed the active zone.

In the case of a homogeneous and linear subsoil, the above conditions are satisfactory. If the geometrical dimensions of the prototype are larger than those of the model by a factor λ, then under an equal load q, the settlement of the prototype is also larger by the factor λ, i.e. the settlement is in direct proportion to the width of the foundation. Such results were obtained, for example, by Dovnarovich and Pol'shin (1967) for medium dense sand, using models of circular foundations with $D = 10$ cm to 60 cm and $q = 0.01$ MPa. Similar results were also obtained by the writer in tests of foundation models with a side of 2.9 to 8 cm (Feda, 1960).

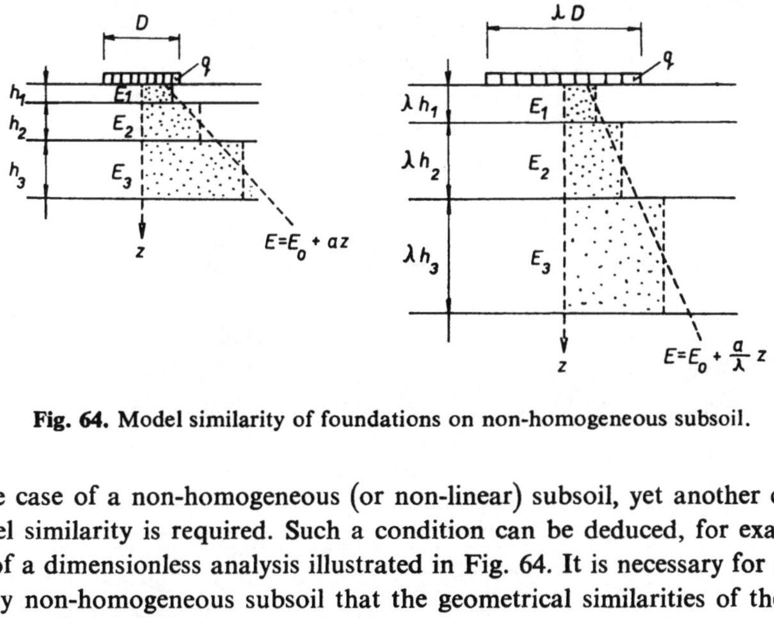

Fig. 64. Model similarity of foundations on non-homogeneous subsoil.

In the case of a non-homogeneous (or non-linear) subsoil, yet another condition of model similarity is required. Such a condition can be deduced, for example, by means of a dimensionless analysis illustrated in Fig. 64. It is necessary for a discontinuously non-homogeneous subsoil that the geometrical similarities of the various layers of the subsoil, in the limiting case, of a continuously non-homogeneous subsoil, satisfy the relation

$$E = E_0 + \frac{a}{\lambda} z \tag{130}$$

where λ is the scale of the reduction of the model. In the form of a dimensionless argument, eqn. (130) may be written as

$$\frac{E}{aD} = \text{const.} \tag{131}$$

and if E = const., then

$$aD = \text{const.} \tag{132}$$

Since by eqn. (62), a is in direct proportion to γ_0, eqn. (132) can be written as

$$\gamma_0 D = \text{const.} \tag{133}$$

Hence, when modelling a linearly and continuously non-homogeneous subsoil one must increase the unit weight of the model subsoil by the factor λ. This condition cannot be satisfied except in a centrifuge, and a fast consolidating medium (sand) should be used for the modelling.

Modelling of this type which is necessary according to what has been said above, was already applied with some success (Pohlshin et al., 1973). The main problem is, of course, the construction of the model subsoil which should be physically identical with the natural subsoil. Undisturbed blocks of the original subsoil can only be used when the subsoil is homogeneous.

Loading plate tests which are frequently performed satisfy neither eqn. (133) nor the relations of model similarity indicated in Fig. 64. If the plates are large enough, the tests demonstrate the non-homogeneity of the subsoil (Fig. 59, the curves a, b), but more often than not they cannot be expected to provide even qualitatively valid predictions of the prototype settlement.

3.12 Effect of subsoil nature on the choice of the method of settlement calculation

Since most methods of settlement calculations are general, they can be applied to any subsoil. An exception are the empirical methods based on penetration tests, which are mainly used for sandy subsoils. Although they can hardly claim generality, some notes concerning this subject may be found useful.

An important criterion for the choice of a particular method of settlement calculations is whether good-quality undisturbed samples can be obtained from the foundation soil under consideration or whether one must rely only on field tests. As mentioned in Section 3.9, some methods of settlement calculations call for laboratory tests.

Engineers depend on field tests in the case of sandy subsoils, of greater depths in stiff to firm clays (at lesser depths, one can use block samples) and in the case of rocks. In the first case, the settlement is usually calculated empirically (although Cornforth, 1974b, points out a sound possibility of applying laboratory results); in the other two cases, the deformation properties of the subsoil are determined by means of plate loading tests or a pressuremeter, and one can apply the theory of isotropic non-homogeneous halfspace (choosing the value of μ).

Another matter of importance is the subsoil non-homogeneity which is influenced by numerous factors, such as the geological structure of the terrain, etc. In the case of a comparatively thin (relative to the foundation dimensions) compressible layer, the deformation of the subsoil approaches uniaxial deformation, etc.

Interesting interpretation possibilities are offered by the theory of continuously non-homogeneous halfspace. A question arises whether the nature of the subsoil affects its non-homogeneity just as it seems to affect its anisotropy (cf. Section 2.4).

Assume (together with Ohde, 1939) that in the case of uniaxial compression

$$d\sigma_z = E_{\text{oed}}\, d\varepsilon_z \tag{134}$$

and the deformation modulus E_{oed} is a power function of σ_z, viz.

$$E_{\text{oed}} = \frac{\sigma_z^k}{F} \tag{135}$$

where F is a constant. For $k = 0$, $E_{\text{oed}} = \text{const.}$, and this is the case of elasticity. Since, however, the structure of soils experiences changes during the deformation process (brought about by the effect of the plastic component of deformation), it is generally $k > 0$. Two cases can be distinguished: $k = 1$ and $k = 0.5$. In the former case the changes of the structure are more intensive and one can, therefore, a priori assume that this case is typical of more compressible cohesive subsoils. Using eqns. (134) and (135) one can deduce for $k = 1$ the relation

$$\varepsilon_z = F \lg \frac{\sigma_z}{\sigma_z^1} + \varepsilon_z^1 \tag{136}$$

(σ_z^1, ε_z^1 – unit stress and strain) which is typical of cohesive soils (cf. eqns. (106) and (113)).

For the same reason the case of $k = 0.5$ can a priori be thought of as belonging to less compressible cohesionless subsoils. For $k = 0.5$, eqns. (134) and (135) give

$$\varepsilon_z = b \left(\frac{\sigma_z}{\sigma_z^1}\right)^{1-k} \tag{137}$$

where b is a parameter. As numerous experimental results prove (cf. for example, Feda, 1976b), for sand compressed in an oedometer, eqn. (137) gives $k = 0.5$ regardless of sand's density (on which the parameter b depends).

Eqns. (136) and (137) and eqn. (91) can be used for settlement calculations. As the results of such calculations show (Sherif, 1973, p. 48), the larger is the value of k, the smaller is the effect of the load q, of the foundation dimension D (or B) and of the unit weight of the subsoil γ_0 on the magnitude of the settlement w. In keeping with this, for example, for a constant q, the effect of the foundation size on w should be smaller in the case of cohesive subsoil. This is proved to be so by Fig. 1 for $q = 0.4$ MPa.

If non-homogeneity of the subsoil is produced by the weight of the soil, $\sigma_z = \gamma_0 z$ should be substituted in eqn. (135). In this case a cohesive foundation soil is more non-homogeneous than a cohesionless soil (m in eqn. (61), if $E = E_t$, is 1 in the first case, and 1/2 in the second). In the simplest case of linear (partial) non-homogeneity, the degree of non-homogeneity can be measured by means of the ratio $E_0/(aD)$, in analogy to eqn. (131).

Calculations of the settlement of foundations resting on subsoils whose non-homogeneity is defined in the above way indicate (Carrier and Christian, 1973; Jamiolkowski, 1974) that the following three regions may be distinguished:

(1) $$\frac{E_0}{aD} \leqq 0.01 \qquad (138)$$

(2) $$\frac{E_0}{aD} \geqq 10 \qquad (139)$$

(3) $$0.01 < \frac{E_0}{aD} < 10 \qquad (140)$$

In the first region the settlement is virtually independent of foundation dimensions; hence Winkler's halfspace can be used (eqn. (94), k_s is dependent on μ). In the second region the effect of non-homogeneity is not distinct enough and one can use the theory of homogèneous halfspace according to which the settlement increases linearly with the dimensions of the loading area. In the third case one must resort to the theory of non-homogeneous halfspace – the settlement approximately increases with the logarithm of the foundation diameter (Carrier and Christian, 1973).

Fig. 1 shows that, for loading areas of 300 to 800 cm², sand behaved like Winkler's halfspace, just as loam did within a very narrow interval in the neighbourhood of 36 cm². From this finding and from eqn. (138) one can deduce that in this case

$$a \doteq 3.5E_0 \text{ cm}^{-1} \qquad (141)$$

in sand, and

$$a \doteq (15 \text{ to } 20)\, E_0 \text{ cm}^{-1} \qquad (142)$$

in loam.

Hence, using the magnitude of a as a criterion, the non-homogeneity of cohesive subsoils is about five times greater than that of cohesionless subsoils in the case being considered. This finding is in agreement with the assumption which is based on eqn. (135) and on the fact that the exponent k is larger in the first case (loam) than in the second. It is, however, hard to imagine that considerations of this sort could be generalized and quantified. All the same, collecting empirical material of this kind may be found to serve a useful purpose.

3.13 Conclusion

As a critical review of the various methods used for settlement calculations suggests, the factor which decides about the reliability of a method is the degree to which one succeeds in introducing the non-linear stress-strain relation and/or the non-homogeneity of a subsoil. For an actual subsoil, the character of this relation depends on the initial (geostatic) stress and on the depth below the surface. To some degree, the effect of depth can be converted to that of the geostatic stress. This conclusion is confirmed to the full by an analysis of the various methods of settlement calculations based on laboratory loading tests (Feda, 1960, 1963). Hence the theory of linear and homogeneous halfspace which starts from the principle of superposition (for an external load, the halfspace is regarded as weightless; the resulting stress field is obtained by superposition of the geostatic stress field upon the stress field produced by an external load) is generally inacceptable unless it is modified to include the non-linearity or the non-homogeneity, as is done, for example, in the stress path method or in the finite element method, or, at least, to bring into use a finite depth of the active zone.

The methods of settlement calculations can be compared from different standpoints. Mathematical rigour need not be the principal criterion, for the ability to provide a physically fitting description of the subsoil, reliability and the amount of time and work necessary to carry out the calculations are all factors that play a decisive role in the appraisal of the suitability of a method.

Physically most realistic solutions are obtained by the application of the stress path method, of the finite element method and of the theory of non-homogeneous halfspace (the non-linear theories are less practical)*). The stress path method predicts settlements on the basis of stress-strain relations at several points (or in several regions) of the subsoil. This simplification endows the stress-path method with a better ability to provide a clear physical picture than does the finite element method; on the other hand, it makes it less rigorous mathematically.

The reliability of a method depends on the degree to which the assumptions on which the method is based are satisfied under field conditions. Although all methods can be successful in this respect, the stress path method and the finite element method are universal, while the applicability of the oedometric compression method is less broad. As the results of the existing research suggest (cf. Section 3.10), greatest emphasis should be put on a truthful description of subsoil non-homogeneity.

As far as the amount of time and work required for carrying out the calculations is concerned, the stress path method and the finite element method are the most demanding of all (the consumption of time and work of the finite element method can

---

*) A homogeneous non-linear halfspace can be replaced by a fictitious linear non-homogeneous halfspace whose non-homogeneity is stress-dependent. The possibility of such an approach was proved by Rogatin's (1976) plate loading test results and discussed in Section 2.7.

greatly be reduced by the introduction of standard programmes). The application of the diagrams shown in Figs. 38 to 40, or the calculation of stresses using Giroud's tables which are appended to this publication, the summation of the strains in the individual layers of the subsoil and correction of the final settlement, taking account of the effect of the depth of foundation according to Fig. 32, are well suited for engineering purposes and comparatively undemanding.

All of the methods described in the foregoing can therefore be said to be applicable to settlement calculations, each with a different degree of generality. The choice of a method depends on the accuracy required of the prediction, the boundary conditions of the problem being considered, the laboriousness of the calculations, the extent of the experimental programme designed for the determination of the in situ mechanical parameters of the subsoil, the nature of the subsoil (whether or not undisturbed samples can be taken – cf. Section 3.9), and the variability of the properties of the subsoil which can – in the extreme case – make every analysis more or less approximate, etc.

The classical methods (in which a gradual introduction of automatic computers is envisaged) will clearly continue to be important for the solution of comparatively simple cases, of problems involving poorly defined boundary conditions, inaccurate or strongly variable mechanical parameters of the subsoil, etc., in which engineering intuition and experience are bound to come into the picture besides other factors. Such conditions and circumstances are, and will be, very common in foundation engineering. The finite element method has increasingly been gaining in importance in solutions of unique problems characterized by the complexity of the subsoil – non-homogeneity and anisotropy – and of the boundary conditions, large scale of the project, high demands on the accuracy of the solution, etc. The method shows undoubtedly good prospects for a future development in applications to foundation soils on condition that it will be oriented towards the use of more realistic constitutive relations, as seems to be its tendency at present. Parametric studies represent an effective use of this method.

Every settlement calculation conceals definite, sometimes considerable, inaccuracies which are independent of the quality of the calculation proper. A mechanical model of the subsoil (i.e. the assignment of realistic mechanical parameters to each point of the subsoil in the region which is important for settlement calculations, the determination of correct boundary conditions, for example, of draining layers, etc.) is never perfect. Its reliability is affected not only by the character of the subsoil but also by the extent and quality of all operations (boreholes, tests) which are necessary for its construction. The geostatic state of stress is never known exactly and the load for which the settlement is calculated may differ considerably from the actual load of stochastical variability. Hence considerations relating to the possibility of successful settlement predictions should differentiate between the suitability and quality of the calculation proper and the effect of these additional, frequently highly important factors.

Future research should go deeper into the problems of settlement in order to enable reliable comparisons to be made between the behaviour of a real subsoil and that of its mathematical model (cf. Section 3.10) and a clear-cut separation of the effects of the calculation method, of the deformation parameters and of their natural as well as artificially produced (parasitic effects accompanying test procedures) variability.

BIBLIOGRAPHY

Agishev, I. A. (1957), Relation between porosity and deformation modulus established by field tests of clayey soils. (Агишев, И. А., 1957, Зависимость между пористостью и модулем деформации установленная полевыми испытаниями глинистых грунтов). Бюлл. Основания и фундаменты, Госстройиздат, № 20, p. 3—6.

Ahlvin, R. G., Ulery, H. H. (1962), Tabulated values for determining the complete pattern of stresses, strains and deflections beneath a uniform load on a homogeneous halfspace. Highway Research Board, Bull. 342, p. 1—13.

Altes, J. (1976) Die Grenztiefe bei Setzungsberechnungen. Der Bauingenieur 51, 3, p. 93—96.

Andrawes, K. Z., El-Sohby, M. A. (1973), Factors affecting coefficient of earth pressure K_0. J. SM Div. ASCE 99, 7, p. 527—539.

Awojobi, A. O. (1975), The settlement of a foundation on Gibson soil of the second kind. Géotechnique 25, 2, p. 221—228.

Awojobi, A. O., Gibson, R. E. (1973), Plane strain and axially symmetric problems of a linearly non-homogeneous elastic half-space. Quart. J. Mech. and Appl. Math. 26, 3, p. 285—302.

Baguelin, F., Jézéquel, J.-F., Le Méhauté, A. (1975), A propos de l'état initial en mécanique des sols. Bull. de liaison L.P.C., 75, p. 148—156.

Barden, L. (1963), Stresses and displacements in a cross-anisotropic soil. Géotechnique 13, 3, p. 198—210.

Bartolomey, A. A. (1975), Results of studies of the distribution of stresses and strains in soil in the active zone of pile foundations. (Бартоломей, А. А., 1975, Результаты исследования распределения напряжений и деформаций в активной зоне свайных фундаментов). Proc. First Baltic COSMFE, Gdańsk, 3, p. 17—28.

Bassett, R. H. (1970), Discussion. Proc. I.C.E. Conf. "Behaviour of Piles", London, p. 208—210.

Bažant, Z., Jun. (1938), Groundwater flow and its effect on the design of foundations of structures, particularly weirs (in Czech). Mas. akad. práce, Praha, 107 pp.

Belloni, L., Jamiolkowski, M. (1973), Sull'impiego del modello di semispazio eterogeneo negli studi di interazione tra struttura di fondazione e terreno. Rivista Italiana di Geotecnica 7, 4, p. 233—249.

Belotti, R., Formigoni, G., Jamiolkowski, M. B. (1975), Remarks on the effect of overconsolidation on the coefficient of earth pressure at rest. Proc. First Baltic COSMFE, Gdańsk, 2, p. 59—71.

Bergfelt, A. (1956), Loading tests on clay. Géotechnique 6, 1, p. 15—31.

Biarez, J. (1962), Contribution à l'étude des propriétés mécaniques des sols et des matériaux pulvérulents. Thèse — Univ. de Grenoble, 240 pp.

Biarez, J., Boulon, M. (1971), Contraintes et déplacements en milieu homogène à loi de comportement non linéaire et en milieu hétérogène à loi élastique linéaire. Proc. 2nd Geot. Conf., Bucarest, 1, p. 65—90.

BIOT, M. (1936), Effet de certaines discontinuités du sous-sol sur la répartition des pressions dues à une charge. Travaux, No. 41.

BISHOP, A. W. (1958), Test requirement for measuring the coefficient of earth pressure at rest. Brussels Conf. 58 on Earth Pressure Problems, 1, p. 2—14.

BISHOP, A. W. (1961), The measurement of pore pressure in the triaxial test. Proc. Conf. "Pore Pressure and Suction in Soils", London, p. 38—46.

BISHOP, A. W., KENNARD, M. F., PENMAN, A. D. M. (1961), Pore pressure observation at Selset dam. Proc. Conf. "Pore Pressure and Suction in Soils", London, p. 91—102.

BISHOP, A. W., WEBB, D. L., LEWIN, P. I. (1965), Undisturbed samples of London clay from the Ashford common shaft; strength-effective stress relationships. Géotechnique 15, 1, p. 1—31.

BJERRUM, L. (1967), Engineering geology of normally consolidated marine clays as related to the settlement of buildings. Seventh Rankine lecture, Géotechnique 17, 2, p. 83—118.

BJERRUM, L. (1973), Problems of soil mechanics and construction in soft clays and structurally unstable soils (collapsible, expansive and other). Proc. 8th ICOSMFE, Moscow, 3, p. 111—159.

BJERRUM, L., ANDERSEN, K. H. (1972), In-situ measurement of lateral pressures in clay. Proc. 5th ECOSMFE, Madrid, 1, p. 11—20.

BJERRUM, L., EGGESTAD, Å. (1963), Interpretation of loading test on sand. Proc. ECOSMFE "Problems of Settlements and Compressibility of Soils", Wiesbaden, 1, p. 199—203.

BJERRUM, L., NASH, J. K. T. L., KENNARD, R. M., GIBSON, R. E. (1972), Hydraulic fracturing in field permeability testing. Géotechnique 22, 2, p. 319—332.

BOLEŃSKI, M. (1973), Settlement of newly constructed buildings in relation to foundation subsoil (in Polish). Inst. techniki budowlanej, Warsaw, 119 pp.

BOLSHTIANSKIY, M. P. (1962), Experimental investigation of stresses in a non-homogeneous cohesionless medium (Болштянский, М. П., 1962, Экспериментальное исследование напряжений в неоднородной сыпучей среде). Осн., фунд. и мех. грунтов №6.

BOND, D. (1961), The influence of foundation size on settlement. Géotechnique 11, 2, p. 121—143.

BOROWICKA, H. (1943), Die Druckausbreitung im Halbraum bei linear zunehmendem Elastizitätmodul. Ingenieur-Archiv 14, 2, p. 75—82.

BOUSSINESQ, J. (1885), Application des potentiels à l'étude de l'équilibre et du mouvement des solides élastiques. Gauthier-Villars, Paris, 721 pp.

BOWLES, J. E. (1968), Foundation analysis and design. McGraw Hill, New York, 659 pp.

BRETH, H., CHAMBOSSE, G. (1974), Discussion. Proc. Conf. "Settlement of Structures", Cambridge, BGS, London 1975, p. 678—681.

BRETH, H., SCHUSTER, E., PISE, P. (1973), Axial stress-strain characteristics of sand. J. SM Div. ASCE 99, 8, p. 617—632.

BRINCH HANSEN, J. (1966), Improved settlement calculation for sand. D. G. I. Bull. 20, p. 15—20.

BROMS, B. B. (1971), Lateral earth pressure due to compaction of cohesionless soils. Proc. 4th COSMFE, Budapest, 1, p. 373—384.

BROOKER, E. W., IRELAND, H. O. (1965), Earth pressure at rest related to stress history. Can. Geot. Journal 2, 1, p. 1—15.

BROWN, P. T., GIBSON, R. E. (1972), Surface settlement of a deep elastic stratum whose modulus increases linearly with depth. Can. Geot. J. 9, 4, p. 467—476.

BROWN, P. T., GIBSON, R. E. (1973), Rectangular loads on inhomogeneous elastic soil. J. SM Div. ASCE 99, 10, p. 917—920.

BROWN, S. F. (1969), Discussion. Proc. Conf. "In Situ Investigations in Soils and Rocks". BGS, London 1970, p. 243—245.

BROWN, S. F. (1973), The measurement of in situ stress and strain in soils. Proc. Conf. "Field Instrumentation in Geotechnical Engineering", BGS, London 1974, p. 38—51.

BRU, J. P., BAGUELIN, F., GOULET, G., JAECK, G., JÉZÉQUEL, J. (1973), Prévision de tassement au pressiomètre et constatations. Proc. 8th ICOSMFE, Moscow, 1.3, p. 25—31.

BURLAND, J. B. (1969a), Deformation of soft clay beneath loaded areas. Proc. 7th ICOSMFE, Mexico, 1, p. 55—63.

BURLAND, J. B. (1969b), Discussion. Proc. 7th ICOSMFE, Mexico, 3, p. 248—250.

BURLAND, J. B. (1969c), Author's replies. Proc. Conf. "In Situ Investigations in Soils and Rocks", BGS, London 1970, p. 61—62.

BURLAND, J. B. (1971), A method of estimating the pore pressures and displacements beneath embankments on soft natural clay deposits. Proc. Roscoe Mem. Symp. "Stress-Strain Behaviour of Soils", Cambridge, p. 505—536.

BURLAND, J. B. (1973), Discussion. Proc. 8th ICOSMFE, Moscow, 4.2, p. 12—13.

BURLAND, J. B., BROMS, B. B., DE MELLO, V. F. B. (1977), Behaviour of foundatinns and structures. State-of-the-art report, Proc. 9th ICOSMFE, Tokyo, 2, p. 495—546.

BURLAND, J. B., LORD, J. A. (1969), The load-deformation behaviour of Middle Chalk at Mundford, Norfolk: a comparison between full-scale performance and in situ and laboratory measurements. Proc. Conf. "In Situ Investigations in Soils and Rocks", BGS, London 1970, p. 3.

BURLAND, J. B., SILLS, G. C., GIBSON, R. E. (1973), A field and theoretical study of the influence of non-homogeneity on settlement. Proc. 8th ICOSMFE, Moscow, 1.3, p. 39—46.

BURLAND, J. B., WROTH, C. P. (1974), Settlement of buildings and associated damage. Review paper — session V. Proc. Conf. "Settlement of Structures", BGS, London 1975, p. 611—654.

BURMISTER, D. M. (1943), The theory of stresses and displacements in layered systems and applications to the design of airport runways. Proc. Highway Res. Board, 23, p. 126—144.

BURMISTER, D. M. (1945), The general theory of stresses and displacements in layered soil systems. J. Appl. Phys. 16, 2, p. 89—94; 3, p. 126—127; 5, p. 296—302.

BURMISTER, D. M. (1956), Stress and displacement characteristics of a two-layer rigid base-soil system: influence diagrams and practical applications. Proc. Highway Res. Board, 35, p. 773—814.

BUTLER, F. G. (1974), Heavily over-consolidated clays. Review paper — session III. Proc. Conf. "Settlement of Structures", BGS, London 1975, p. 531—578.

CAMPANELLA, R. G., VAID, Y. P. (1973), Influence of stress path on the plane strain behaviour of a sensitive clay. Proc. 8th ICOSMFE, Moscow, 1.1, p. 85—92.

CAQUOT, A., KÉRISEL, J. (1966), Traité de mécanique des sols. Gauthier-Villars, Paris, 4. ed., 506 pp.

CARILLO, N. (1948), Influence of artesian wells in the sinking of Mexico City. Proc. 2nd ICOSMFE, Rotterdam, 7, p. 156—159.

CARRIER, D. W., CHRISTIAN, J. T. (1973), Rigid circular plate resting on a non-homogeneous elastic half-space. Géotechnique 23, 1, p. 67—84.

CASAGRANDE, A. (1936), The determination of the pre-consolidation load and its practical significance. Proc. ICOSMFE, Cambridge (Mass), 3, p. 60—64.

CERRUTI, V. (1882), Ricerche intorno all'equilibrio de corpi elastici isotropi. Reale Accademia dei lincei, 13, year of publ. 1881—1882.

CHANG, Y. C. E., BROMS, B., PECK, R. B. (1973), Relationship between the settlement of soft clays and excess pore pressure due to imposed loads. Proc. 8th ICOSMFE, Moscow, 1.1, p. 93—96.

CHANG C.-Y., DUNCAN, J. M. (1970), Analysis of soil movement around a deep excavation. J. SM Div. ASCE 96, 5, p. 1655—1681.

CHIN, F. K. (1969), Size and load effect on settlement of footings in clay. Proc. 7th ICOSMFE, Mexico, 2, p. 53—58.

CHRISTIAN, J. T., WONG, I. H. (1973), Errors in simulating excavation in elastic media by finite elements. Soils and Foundations 13, 1, p. 1—10.

CORNFORTH, D. H. (1974a), Discussion. Proc. Conf. "Settlement of Structures", BGS, London 1975, p. 759—760.

CORNFORTH, D. H. (1974b), Discussion. Proc. Conf. "Settlement of Structures", BGS, London 1975, p. 681—682.
CORNFORTH, D. H. (1974c), One dimensional consolidation curves of a medium sand. Géotechnique 24, 4, p. 678—683.
COZZOLINO, E. V. M. (1961), Statistical forecasting of compression index. Proc. 5th ICOSMFE, Paris, 1, p. 51—53.
CRAWFORD, C. B. (1964), Interpretation of the consolidation test. J. SM Div. ASCE 90, 5, p. 87—102.
CRAWFORD, C. B., EDEN, W. Y. (1965), A comparison of laboratory results with in-situ properties of Leda clay. Proc. 6th ICOSMFE, Montreal, 1, p. 31—35.
ČERMÁK, Z., KLAPETEK, F. (1959), A comparison of calculated and actual settlement of massive hydraulic structures (in Czech). Inž. stavby, 7, 12, p. 464—470.
DALMATOV, B. I., SOTNIKOV, S. N., DOROSHKEVICH, N. M., ZNAMENSKY, V. V. (1973), Investigation of soil deformation in foundation beds of structures. Proc. 8th ICOSMFE, Moscow, 1.3, p. 55—60.
D'APPOLONIA, D. J., D'APPOLONIA, E. E., BRISSETTE, R. F. (1968), Settlement of spread footings on sand. J. SM Div. ASCE 94, 3, p. 735—760.
D'APPOLONIA, D. J., D'APPOLONIA E. E., BRISSETTE, R. F. (1970), Discussion on settlement of spread footings on sand. J. SM Div. ASCE 96, 2, p. 754—761.
D'APPOLONIA, D. J., LAMBE, T. W. (1970), Method for predicting initial settlement. J. SM Div. ASCE 96, 2, p. 523—545.
D'APPOLONIA, D. J., POULOS, H. G., LADD, CH. C. (1971), Initial settlement of structures on clay. J. SM Div. ASCE 97, 10, p. 1359—1377.
D'APPOLONIA, D. J., LAMBE, T. W., POULOS, H. G. (1971), Evaluation of pore pressure beneath an embankment. J. SM Div. ASCE 97, 6, p. 881—897.
DAVIS, E. H., POULOS, H. G. (1963), Triaxial testing and three-dimensional settlement analysis. Proc. 4th Australia-New Zealand COSMFE, p. 233.
DAVIS, E. H., POULOS, H. G. (1968), The use of elastic theory for settlement prediction under three-dimensional conditions. Géotechnique 18, 1, p. 67—91.
DE BEER, E. (1965), Bearing capacity and settlement of shallow foundations on sand. Proc. Symp., "Bearing Capacity and Settlement of Foundations", Duke Univ. (publ. 1967), p. 15—33.
DE BEER, E. (1972), Heutiger Stand der Berechnung der Sohldruckverteilung unter Flächengründungen bei besonderer Berücksichtigung der Bauerwerkssteifigkeit. Proc. Inter. Symp. Dresden, p. 23—39.
DESAI, C. S. (1971), Nonlinear analyses using spline functions. J. SM Div. ASCE 97, 10, p. 1461—1480.
DESAI, C. S. (1975), Soil-structure interaction and simulation problems with some related topics. Int. Symp. on Num. Methods in Soil Mech. and Rock Mech., Karlsruhe — preprint, 71 pp. (in; Gudehus, G., ed., Finite Elements in Geomechanics, Wiley & Sons, 1977, p. 209—250).
DESAI, C. S., ABEL, J. F. (1972), Introduction to the finite element method. Van Nostrand Reinhold Co., New York, 477 pp.
DIES, DUPAS, J. M., LONDEZ, M. (1973), Un problème de déformation — méthodes et résultats. Proc. 8th ICOSMFE, Moscow, 1.1, p. 97—101.
DOLEŽALOVÁ, M. (1974), Geostatic stress state in cross-anisotropic soil deposits. Proc. 4th Danube-European COSMFE, Bled, p. 155—160.
DOLEŽALOVÁ, M., TRUNCOVÁ, D., VORLÍČKOVÁ, Z. (1975), In situ determination of initial state of stress in a rock mass and the finite element method (in Czech). Res. Rep. Hydroprojekt Praha, 21 pp. + 11 encl.
DOOLEY, J. C. (1964), Correspondence. Géotechnique 14, 3, p. 278—279.
DOVNAROVICH, S. V. (1975), Deformation of soil layers of different thicknesses on a rigid base.

(Довнарович, С. В., 1975, Деформирование жесткоподстилаемых слоев грунта при разной толщине слоя). Осн., фунд. и мех. грунтов, 5, p. 40—42.

Dovnarovich, S. V., Polshin, D. E. (1967), On the choice of the dimensions of foundation models in the modelling of settlements of a sandy subsoil. (Довнарович, С. В., Польшин, Д. Е., 1967, О выборе размера модели фундамента при моделировании осадок песчаного основания). Осн., фунд. и мех. грунтов, 4, p. 27—28.

Duffy, J. (1959), A differential stress-strain relation for the hexagonal close-packed array of elastic spheres. J. Appl. Mech. 26, 1, p. 88—94.

Duffy, J., Mindlin, R. D. (1957), Stress-strain relations and vibrations of a granular medium. J. Appl. Mech. 24, 4, p. 585—594.

Duncan, J. M., Chang Ch.-Y. (1970), Nonlinear analysis of stress and strain in soils. J. SM Div. ASCE 96, 5, p. 1629—1653.

Duncan, J. M. (1973), Discussion. Proc. 8th ICOSMFE, Moscow, 4.2, p. 17.

Duncan, J. M., Monismith, C. L., Wilson, E. L. (1968), Finite element analysis of pavements. Highway Res. Board, vol. 228.

Dunn, C. S. (1974a), Settlement of a large raft foundation on sand. Proc. Conf. "Settlement of Structures", BGS, London 1975, p. 14—21.

Dunn, C. S. (1974b), Discussion. Proc. Conf. "Settlement of Structures", BGS, London 1975, p. 676—678.

Dzwilewski, P. T., Richards, A. F. (1974), Consolidation properties of Wilkinson basin soils. J. Geot. Eng. Div. 100, 10, p. 1175—1179.

Eggestad, A. (1963), Deformation measurements below a model footing on the surface of dry sand. Proc. 3rd ECOSMFE "Problems of Settlements and Compressibility of Soils", Wiesbaden, 1, p. 233—239.

Egorov, K. E. (1959), Determination of the deformations of different subsoil layers beneath a chimney. (Егоров, К. Е., 1959, Изучение послойной деформации основания дымовой трубы). Осн., фунд. и мех. грунтов, 4, p. 4—7.

Egorov, K. E., Popova, O. V. (1971a), Comparison of computed and factual settlements of high smoke stacks foundations. Proc. 4th Asian Reg. COSMFE, Bangkok, 1, p. 9—15.

Egorov, K. E., Popova, O. V. (1971b), Settlement of foundations of tall structures. (Егоров, К. Е., Попова, О. В., 1971b, Осадки фундаментов сооружений башенного типа). Сб. конф. ,,Основания, фундаменты и механика грунтов", Киев, p. 297—302.

Esu, F., Calabresi, G. (1969), Slope stability in an overconsolidated clay. Proc. 7th ICOSMFE, Mexico, 2, p. 555—563.

Feda, J. (1960), Limit load and deformation of a cohensionless medium (in Czech). Res. Rep. ÚTAM ČSAV, Praha, 180 pp. + 56 figs.

Feda, J. (1963), Validity of some settlement computation theories as tested in laboratory conditions on granular soils. Proc. 3rd ECOSMFE "Problems of Settlements and Compressibility of Soils", Wiesbaden, 1, p. 61—67.

Feda, J. (1970), Heaving of cohesive foundation soil by the effect of swelling (in Czech). Inž. stavby 18, 9, p. 338—343.

Feda, J. (1971a), Discussion on a paper by V. Mach: "Laboratory studies ..." (in Czech). Stav. časopis 19, 9, p. 711—716.

Feda, J. (1971b), Discussion on "Effect of strain history on liquefaction of sand". J. SM Div. ASCE 97, 6, p. 945—950.

Feda, J. (1972), Time-independent constitutive relations of soils (in Czech). Res. Rep. ÚTAM ČSAV, Praha, 112 pp. + 17 figs.

Feda, J. (1976a), Theory of pressure at rest of cohesionless soils (in Czech). Stav. čas. 24, 9—10, p. 729—749.

Feda, J. (1976b), Uniaxial compressibility of Zbraslav sand (in Czech). Stav. čas. 24, 8, p. 614—639.

FEDA, J. (1976c), The effect of the subsoil heterogeneity on the calculation of settlement of structures (in Czech). Inž. stavby 24, 7, p. 337—343.

FEDA, J., HAVLÍČEK, J., SEYČEK, J., ŠKOPEK, J. (1977), Active zone beneath foundations. Proc. 9th ICOSMFE, Tokyo, 1, p, 505—510.

FILON, L. N. G. (1904), Phil. Trans. Roy. Soc., London.

FLORIN, V. A. (1959), Fundamentals of soil mechanics. (Флорин, В. А., 1959, Основы механики грунтов. Том 1). Госстройиздат, Москва—Ленинград, 357 pp.

FLORIN, V. A. (1961), Fundamentals of soil mechanics. (Флорин, В. А., 1961, Основы механики грунтов. Том 2). Госстройиздат, Москва—Ленинград, 543 pp.

FOLAYAN, J. I., HOËG, K., BENJAMIN, J. R. (1970), Decision theory applied to settlement predictions. J. SM Div. ASCE 96, 4, p. 1127—1141.

FOX, L. (1948a), Computations of traffic stresses in a simple road structure. Proc. 2nd ICOSMFE, Rotterdam, 2, p. 236—246.

FOX, E. N. (1948b), The mean elastic settlement of a uniformly loaded area at a depth below the ground surface. Proc. 2nd ICOSMFE, Rotterdam, 1, p. 129—132.

FRONTARD, I. (1914), Notice sur l'accident de la digue de Chartres. Ann. des Ponts et Chaussées 9, 23, p. 173—280.

FRÖHLICH, O. K. (1934), Druckverteilung im Baugrunde. J. Springer, Wien, 185 pp.

GARGA, V. K., QUIN, J. T. (1974), An investigation on settlements of direct foundations on sand. Proc. Conf. "Settlement and Structures", BGS, London 1975, p. 22—36.

GASSMAN, F. (1951, 1953), Elastic waves through a packing of spheres. Geophysics 16, p. 673—685 (1951); 18, p. 268 (1953).

GAZIEV, E. G. (1973), Rock mechanics in civil engineering. (Газиев, Е. Г., 1973, Механика скальных пород в строительстве). Стройиздат, Москва, 177 pp.

GEDDES, J. D. (1966), Stresses in foundation soils due to vertical subsurface loading. Géotechnique 16, 3, p. 231—255.

GERRARD, C. M. (1969), Tables of stresses, strains and displacements in two-layer elastic systems under various traffic loads. Austr. Road Res. Board, Spec. Rep. No. 3, 157 pp.

GERRARD, C. M. (1975), Background to mathematical modelling in geomechanics — the roles of fabric and stress history. Int. Symp. Num. Methods in Soil Mech. and Rock Mech., Karlsruhe; preprint, 52 + VII pp., 40 figs., 1 table. (in: Gudehus, G., ed., Finite Elements in Geomechanics, Wiley & Sons, 1977, p. 33—120).

GERRARD, C. M., HARRISON, W. J. (1970a), Circular loads applied to a cross anisotropic halfspace. CSIRO — Div. of Appl. Geom., Tech. paper No. 8, 61 pp.

GERRARD, C. M., HARRISON, W. J. (1970b), Stresses and displacements in a loaded orthorhombic halfspace. CSIRO — Div. of Appl. Geom., Tech. paper No. 9, 27 pp.

GERRARD, C. M., MORGAN, J. R. (1969), Application of layered elastic theory to practical problems. Proc. 7th ICOSMFE, Mexico, 1, p. 143—150.

GHABOUSSI, J., WILSON, E. L., ISENBERG, J. (1973), Finite element for rock joints and interfaces. J. SM Div. ASCE 99, 10, p. 833—848.

GIBSON, R. E. (1967), Some results concerning displacements and stresses in a non-homogeneous elastic half-space. Géotechnique 17, 1, p. 58—67.

GIBSON, R. E. (1974), The analytical method in soil mechanics. Géotechnique 24, 2, p. 115—140.

GIBSON, R. E., ANDERSON, W. F. (1961), In situ measurement of soil properties with the pressuremeter. Civ. Eng. Publ. Works Rev. 56, 5, p. 615—620.

GIBSON, R. E., BROWN, P. T., ANDREWS, K. R. F. (1971), Some results concerning displacements in a non-homogeneous elastic layer. Zeit. f. angew. Math. u. Phys. 22, 5, p. 855—864.

GIBSON, R. E., MARSLAND, A. (1961), Pore-water pressure observations in a saturated alluvial deposit beneath a loaded oil tank. Proc. Conf. "Pore Pressure and Suction in Soils", London, p. 112—118.

Gibson, R. E., Sills, G. C. (1971), Some results concerning the plane deformation of a non-homogeneous elastic half-space. Proc. Roscoe Mem. Symp. "Stress-Strain Behaviour of Soils", Cambridge, p. 564—572.

Gielly, J., Lareal, P., Sanglerat, G. (1969), Correlations between in situ penetrometer tests and the compressibility characteristics of soils. Proc. Conf. "In Situ Investigations in Soils and Rocks", BGS, London, p. 167—172.

Girijavallabhan, Ch. V., Reese, L. C. (1968), Finite-element method for problems in soil mechanics. J. SM Div. ASCE 94, 2, p. 473—496.

Giroud, J.-P. (1968), Settlement of a linearly loaded rectangular area. J. SM Div. ASCE 94, 4, p. 813—831.

Giroud, J.-P. (1969), Fondation rectangulaire linéairement chargée: tassement et contraintes. Ann. de l'ITBTP, No. 253, p. 81—111.

Giroud, J.-P. (1970a), Stresses under a linearly loaded rectangular area. J. SM Div. ASCE 96, 1, p. 263—268.

Giroud, J.-P. (1970b), Commentaires sur l'utilisation de l'élasticité pour les projects de fondation. Proc. 4th Congr. Brasil Soc. Soil Mech. Found. Eng., Rio de Janeiro, 1, 2, p. 5-1—5-9 (cf. Technique des Travaux, 331, p. 264—266, 1971).

Giroud, J.-P. (1972), Tables pour le calcul des fondations. Tome 1 — Tassement, Dunod, 360 pp.

Giroud, J.-P. (1973), Tables pour le calcul des fondations. Tome 2 — Tassement, Dunod, 505 pp.

Giroud, J.-P., Cordary, D. (1975), Estimation et correction des approximations classiques du calcul du tassement des fondations superficielles. Ann. de l'ITBTP, No. 333, p. 105—131.

Giroud, J.-P., Watissee, H., Rabatel, A. (1970), Tassements et contraintes dans une couche de sol élastique supportant une charge uniformément répartie. Bull. Liason Lab. Routiers, Ponts et Chaussées, 48, p. 97—124.

Gol'dstein, M. N. (1971), On some problems of present-day soil mechanics and foundation engineering. (Гольдштейн, М. Н., 1971, О некоторых проблемах современной механики грунтов и фундаментостроения). Осн., фунд. и мех. грунтов, 2, p. 7—10.

Goodman, R. E., Dubois, J. (1972), Duplication of dilatancy in analysis of jointed rocks. J. SM Div. ASCE 98, 4, p. 399—422.

Goodman, R. E., Taylor, R. L., Brekke, T. L. (1968), A model for the mechanics of jointed rocks. J. SM Div. ASCE 94, 3, p. 637—659.

Gorbunov-Posadov, M. I. (1972), Solution of the problem of point load applied to the boundary of a halfspace with displacements attenuating in the infinity. (Горбунов-Посадов, М. И., 1972, Решение задачи о сосредоточенной силе, приложенной к границе полуплоскости, при условии затухания перемещений на бесконечности). Осн., фунд., и мех. грунтов, 2, p. 2—6.

Gorbunov-Posadov, M. I., Malikova, T. A. (1973), Calculations of structures on an elastic base. (Горбунов-Посадов, М. И., Маликова, Т. А., 1973, Расчет конструкций на упругом основании). Госстройиздат, Москва, 627 pp.

Green, G. E. (1971), Strength and deformation of sand measured in an independent stress control cell. Proc. Roscoe Mem. Symp. "Stress-Strain Behaviour of Soils", Cambridge, p. 285—323.

Griffith, J. H. (1929), The pressures under substructures. Iowa State Univ. Eng. Exp. Sta. Bull. 117, and Engineering and Constructions, 1, p. 113—119.

Hanshaw, B. B., Bredehoeft, J. D. (1968), On the maintenance of anomalous fluid pressures. I. and II. Bull. 70, Geol. Soc. of America.

Harrison, W. J., Wardle, J. L., Gerrard, C. M. (1972), Computer programmes for circle and strip loads on layered anisotropic media. Div. of Appl. Geomechanics, CSIRO, Australia, 131 pp.

Hendron, A. J. (1963), Behaviour of sand in one-dimensional compression. Ph. D. thesis, Univ. of Illinois, 285 pp.

HENKEL, D. J. (1960), The shear strength of saturated remolded clay. Proc. Res. Conf. Shear Strength Cohes. Soils, ASCE, p. 533—554.

HLAVÁČEK, J. (1975), Settlement of pile foundation surroundings. Proc. First Baltic COSMFE, Gdańsk, 3, p. 95—105.

HOËG, K., CHRISTIAN, J. T., WHITMAN, R. V. (1968), Settlement of a strip footing on elastic-plastic soil. J. SM Div. ASCE 94, 2, p. 431—445.

HOËG, K., ANDERSLAND, O. B., ROLFSEN, E. N. (1969), Undrained behaviour of quick clay under load tests at Asrum. Géotechnique 19, 1, p. 101—115.

HOBBS, N. B. (1974), Factors affecting the prediction of settlement of structures on rock: With particular reference to the chalk and Trias. Review paper — session IV. Proc. Conf. "Settlement of Structures", BGS, London 1975, p. 579—610.

HOBBS, N. B., DIXON, J. C. (1969), In-situ testing for bridge foundations in the Devonian Marl. Proc. Conf. "In Situ Investigations in Soils and Rocks", BGS, London 1970, p. 31—38.

HOLL, D. (1940), Stress transmission in earths. Proc. Highway Res. Board, 20, p. 709—721.

HOWARD, J. (1966), Vertical normal stress in the earth and the weight of overburden. Geol. Soc. America, Bull. 5, 77, p. 651—660.

HRUBAN K. (1945), Strains and stresses in a non-homogeneous elastic halfspace (in Czech). Rozpravy II. tř. České akademie věd a umění, vol. 55, No. 13, 23 pp.

HRUBAN, K. (1948), Sources of error in settlement estimate. Proc. 2nd ICOSMFE, Rotterdam, 1, p. 127—128.

HRUBAN, K. (1958), Strains and stresses in a physically non-linear halfspace (in Czech). Rozpravy ČSAV, řada TV, vol. 68, No. 1, 52 pp.

HUANG, Y. H. (1968), Stresses and displacements in nonlinear soil media. J. SM Div. ASCE 94, 1, p. 1—19.

HUDER, J., SEVALDSON, R. A. (1963), Setzungen und Verfestigung im strukturempfindlichen Untergrund. Proc. ECOSMFE "Problems of Settlements and Compressibility of Soils", Wiesbaden, 1, p. 337—341.

IGNATOVA, O. I. (1968), On the correction of the oedometric deformation modulus of clayey soils of plastic consistency (Игнатова, О. И., 1968, Корректировка значений модулей деформации глинистых грунтов пластической консистенции, определенных на компресионных приборах). Осн., фунд. и мех. грунтов, 2, p. 8—10.

IVANOV, N. N. (1926), On the situation in technical testing of soils. (Иванов, Н. Н., 1926, К постановке технических испытаний грунтов). Сб. ЦУМТ, Трансиздат.

JÁKY, J. (1944), The coefficient of earth pressure at rest (in Hungarian). A Magyar Mérnök- és Építész-Egylet Közlönye, 78, 22, p. 355—358.

JAMIOLKOWSKI, M. B. (1974), Discussion. Proc. Conf. "Settlement of Structures", BGS, London 1975, p. 785—787.

JANBU, N., BJERRUM, L., KJAERNSLI, B. (1956), Soil mechanics applied to some engineering problems (in Norwegian). Publ. N.G.I., No. 16, 93 pp.

JELINEK, R. (1951), Der Einfluss der Gründungstiefe und begrenzter Schichtmächtigkeit auf die Druckausbreitung im Baugrund. Die Bautechnik 28, 6, p. 125—130.

JÉZÉQUEL, J. F. (1973), Discussion — specialty session 5. Proc. 8th ICOSMFE, Moscow, 4.3, p. 243—247.

JONES, A. (1962), Tables of stresses in three-layer elastic system. Highway Res. Board, Bull. 342, p. 176—214.

KAFKA, V. (1964), Solution of stresses in a periodically non-homogeneous layered medium (in Czech). Rozpravy ČSAV — TV 74, 4, 78 pp.

KANDAUROV, I. I. (1959), Theory of discrete distribution of vertical stresses and compression deformations in homogeneous and multilayered subsoils. (Кандауров, И. И., 1959, Теория дискретного распределения вертикальных напряжений и деформаций сжатия в однородных и многослойных грунтовых основаниях). ВАТТ Москва.

KANDAUROV, I. I. (1960), Theory of stress distribution in a granular subsoil. (Кандауров, И. И., 1960, К теории распределения напряжений в зернистом грунтовом основании). Осн., фунд. и мех. грунтов № 4.

KANDAUROV, I. I. (1966), Mechanics of granular media and its application to civil engineering. (Кандауров, И. И., 1966, Механика зернистых сред и ее применение в строительстве). Госстройиздат, Москва, 319 pp.

KANDAUROV, I. I., AKSENOV, E. M., MATVEEV, V. P. (1969), Calculation of the state of stress and settlement of subsoils performed by means of digital computers. (Кандауров, И. И., Аксенов, Е. М., Матвеев, В. П., 1969, Расчет напряженного состояния и осадок оснований с применением цифровых вычислительных машин). Стройиздат, Ленинград, 191 pp.

KENNEY, T. C. (1967), Field measurement of in situ stresses in quick clay. Proc. Geot. Conf. Oslo, 1, p. 49—55.

KENNEY, T. C., MOUM, J., BERRE, T. (1967), An experimental study of bonds in natural clay. Proc. Geot. Conf. Oslo, 1, p. 65—69.

KÉZDI, Á. (1952), Einige Probleme der Spannungsverteilung im Boden. Acta Tech. Acad. Sci. Hung. 2, 2—4, p. 371—389.

KÉZDI, Á. (1962), Erddrucktheorien. Springer, Berlin, 319 pp.

KÉZDI, Á. (1964), Bodenmechanik. Band 2. Verl. f. Bauw., Berlin, 459 pp.

KÉZDI, Á. (1974), Soil physics. Handbook of soil mechanics vol. 1. Akadémiai Kiadó, Budapest, 292 pp.

KLEIN, G. K. (1956), Calculation of structures on a continuous subsoil taking into account non-homogeneity, discontinuity of deformations and other mechanical properties. (Клейн, Г. К., 1956, Учет неоднородности, разрывности деформаций и других механических свойств грунта при расчете сооружений на сплошном основании). Сб. трудов МИСИ № 14, Госстройиздат, p. 168—180.

KOLÁŘ, V., KRATOCHVÍL, J., ZLÁMAL, M., ŽENÍŠEK, A. (1971), Technical, physical and mathematical principles of the finite element method. Rozpravy ČSAV-TV 81, 2, 85 pp.

KONDNER, R. (1962), Friction pile groups in cohesive soils. J. SM Div. ASCE 88, 3, p. 117—149.

KONING, H. (1957), Stress distribution in a homogeneous, anisotropic, elastic semi-infinite solid. Proc. 4th ICOSMFE, London, 1, p. 335—338.

KONOVALOV, P. A. (1971), Experimental investigations of deformations of sandy and clayey subsoils. (Коновалов, П. А., 1971, Экспериментальное исследование деформаций песчаных и глинистых оснований). Proc. 2nd Geot. Conf., Bucarest, 1, p. 281—294.

KONOVALOV, P. A., RUDNITSKIY, N. J. (1964), On the coefficient of variability of the deformation modulus of soils. (Коновалов, П. А., Рудницкий, Н. Я., 1964, О коэффициенте изменчивости модуля деформации грунта). Осн. фунд. и мех. грунтов, 3, p. 16—17.

KÖGLER, F., SCHEIDIG, A. (1928), Druckverteilung im Baugrund. Die Bautechnik 15, p. 205—209.

KÖGLER, F., SCHEIDIG, A. (1938), Baugrund und Bauwerk. W. Ernst & Sohn, Berlin, 288 pp.

KRAVTSOV, G. I. (1970), Stress-strain state of a wetted loess foundation soil beneath a circular rigid plate. (Кравцов, Г. И., 1970, Напряженно-деформированное состояние увлажняемого лессового основания круглого жесткого штампа). Сб. „Вопросы мех. грунтов и строит. на лесс. основаниях", Грозный, p. 108—124.

KRIEGEL, H. J., WIESNER, H. H. (1973), Problems of stress-strain conditions of subsoil. Proc. 8th ICOSMFE, Moscow, 1.3, p. 133—141.

KRYZHANOVSKIY, A. L., CHEVIKIN, A. S., KULIKOV, O. V. (1975), Efficiency of calculations of subsoils considering non-linear deformation properties of soils. (Крыжановский, А. Л., Чевикин, А. С., Куликов, О. В., 1975, Эффективность расчета оснований с учетом нелинейных деформационных свойств грунтов). Осн., фунд. и мех. грунтов, 5, 37—40.

KULCHITSKIY, G. B. (1969), Errors associated with the determination of the deformation modulus of soils. (Кульчицкий, Г. Б., 1969, Об ошибках при определении модуля деформации грунтов). Осн., фунд. и мех. грунтов, 3, p. 20—21.

KULHAWY, F. H., DUNCAN, J. M. (1972), Stresses and movements in Oroville dam. J. SM Div. ASCE 98, 7, p. 653—665.

LADD, C. C. (1964), Stress-strain modulus of clay in undrained shear. J. SM Div. ASCE 90, 5, p. 103—132.

LADD, C. C. (1965), Stress-strain behaviour of anisotropically consolidated clays during undrained shear. Proc. 6th ICOSMFE, Montreal, 1, p. 282—286.

LADD, C. C. (1973), Discussion. Proc. 8th ICOSMFE, Moscow, 4.2, p. 108—115.

LADD, C. C., FOOT, R., ISHIHARA, K., SCHLOSSER, F., POULOS, H. G. (1977), Stress-deformation and strength characteristics. State-of-the-art report, Proc. 9th ICOSMFE, Tokyo, 2, p. 421—494.

LAMBE, T. W. (1963), An earth dam for the storage of fuel oil. Proc. 2nd Panam. COSMFE, São Paulo, 2, p. 257—308.

LAMBE, T. W. (1964), Methods of estimating settlement. J. SM Div. ASCE 90, 5, p. 43—67.

LAMBE, T. W. (1967), Stress path method. J. SM Div. ASCE 93, 6, p. 309—331.

LAMBE, T. W. (1970), The integrated civil engineering project. The seventh Terzaghi lecture. In: Terzaghi lectures 1963—1972, ASCE 1974, p. 314—335.

LAMBE, T. W. (1973), Predictions in soil engineering. Géotechnique 23, 2, p. 149—202.

LAMBE, T. W., WHITMAN, R. V. (1969), Soil Mechanics. J. Wiley & Sons, New York, 553 pp.

LAPKIN, V. B. (1975), Effect of cross anisotropy on the state of stress and strain in subsoil loaded by a strip foundation. (Лапкин, В. Б., 1975, Влияние поперечной анизотропии на напряженно-деформированное состояние основания нагруженного ленточным фундаментом). Осн., фунд. и мех. грунтов, 3, p. 37—39.

LA ROCHELLE, P. (1973), Discussion. Proc. 8th ICOSMFE, Moscow, 4.2, p. 102—108.

LA ROCHELLE, P., LEFEBVRE, G. (1971), Sampling disturbance in Champlain clays. ASTM STP No. 483 "Sampling of Soil and Rock", p. 143—163.

LEONARDS, G. A. (1974), Discussion. Proc. Conf. "Settlement of Structures", BGS, London 1975, p. 669.

LEONARDS, G. A., ALTSCHAEFFL, A. G. (1964), Compressibility of clay. J. SM Div. ASCE 90, 5, p. 133—155.

LEONARDS, L., GIRAULT, P. (1961), A study of the one-dimensional consolidation test. Proc. 5th ICOSMFE, Paris, 1, p. 213—218.

LEUSSINK, H. (1954), Das seitliche Nichtanliegen der Bodenprobe im Kompressionsapparat als Fehlerquelle beim Druck-Setzungsversuch. Fortschritte u. Forschungen im Bauwesen, Heft 17, 153 pp.

LITWINISZYN, J. (1957), Fundamental principles of the mechanics of stochastic media. Proc. 3rd Congr. Theor. Appl. Mech., Bangalore, p. 93—110.

LITWINISZYN, J. (1964), An application of the random walk argument to the mechanics of granular media. Symp. IUTAM "Rheology and Soil Mechanics", Grenoble, Springer 1966, p. 82—89.

LO, K. Y., SEYCHUK, J. L., ADAMS, J. I. (1971), A study of the deformation characteristics of a stiff fissured clay. ASTM STP No. 483 "Sampling of Soil and Rock", p. 60—76.

LOBANOV, I. E. (1965), Effect of subsoil non-homogeneity on the accuracy of calculations of settlement of engineering structures. (Лобанов, И. Е., 1965, Влияние неоднородности грунтового основания на точность расчета осадок инженерных сооружений). Труды Новосиб. инст. инж. ж. д. транспорта, вып. 41, p. 119—133.

LOMISE, G. M., KRYZHANOVSKY, A. L., VORONCOV, E. I., GOLDIN, A. L. (1969), Study on deformation and strength of soils under three-dimensional state of stress. Proc. 7th ICOSMFE, Mexico, 1, p. 257—265.

Lomize, G. M., Kravtsov, G. I. (1969), In-situ investigation of the stress-strain state of a loess subsoil under axisymmetric conditions. (Ломизе, Г. М., Кравцов, Г. И., 1969, Опыт натурного исследования напряженно-деформированного состояния лессового основания в условиях осесимметричной задачи). Осн., фунд. и мех. грунтов, 3, p. 3—6.

Lomize, G. M., Kryzhanovskiy, A. L., Petrianin, V. F. (1972), Investigation of the development of the stress-strain state of a sandy subsoil under plane strain conditions. (Ломизе, Г. М., Крыжановский, А. Л., Петрянин, В. Ф., 1972, Исследование закономерностей развития напряженно-деформированного состояния песчаного основания при плоской деформации). Осн., фунд. и мех. грунтов, 1, p. 4—7.

Love, A. E. H. (1927), A treatise on the mathematical theory of elasticity. Cambridge Press, 4th ed., 643 pp.

Lur'ie, A. I. (1955), Three-dimensional problems of elasticity theory. (Лурье, А. И., 1955, Пространственные задачи теории упругости). Гостехтеориздат, Москва.

Malikova, T. A. (1972), Analysis of natural settlements of plate and box foundations of multistorey buildings. (Маликова, Т. А., 1972, Анализ натурных осадок плитных и коробчатых фундаментов многоэтажных зданий). Осн., фунд. и мех. грунтов, 2, p. 17—21.

Malina, H. (1969), Berechnung von Spannungsumlagerungen in Fels und Boden mit Hilfe der Elementenmethode. Veröffentl. d. Inst. f. Bodenm. u. Felsm. d. Univ. Karlsruhe, No. 40, 89 pp.

Marchetti, S. (1975), Comportamento, progetto e taratura delle celle di pressione a diaframma per usi geotecnici. Rivista Italiana di Geotecnica 9, 2, p. 55—83.

Marguerre, K. (1933), Spannungsverteilung und Wellenausbreitung in der kontinuierlich gestützen Plate. Ingenieur Archiv, 4, 4.

Marivoet, L. (1948), A simplified method for computing the bearing capacity of the soil supporting footings or piers. Proc. 2nd ICOSMFE. Rotterdam, 1, p. 78—86.

Marsal, R. J. (1975), The lacustrine clays of the valley of Mexico. Instituto de Ingenieria Mexico Publ. E 16, 26 pp.

Marsland, A. (1973), Laboratory and in-situ measurements of the deformation moduli of London clay. Bldg. Res. Establ. CP 24/73, 12 pp.

Marsland, A., Eason, B. J. (1973), Measurement of displacements in the ground below loaded plates in deep boreholes. Proc. Conf. "Field Instrumentation in Geotechnical Engineering", BGS, London 1974, p. 304—317.

Mathian, J., Paubel, R. (1969), Mouvements verticaux d'une usine-écluse du Rhône. Proc. 7th ICOSMFE, Mexíco, 2, p. 178—182; 3, p. 265—266.

McGown, A., Barden, L., Lee, S. H. (1974), Sample disturbance in soft alluvial Clyde Estuary clay. Can. Geot. J. 11, 4, p. 651—660.

McNamee, J., Gibson, R. E. (1960), Displacement functions and linear transforms applied to diffusion through porous elastic media. Quart. J. Math. Appl. Mech. 13, p. 98—111.

Meigh, A. C. (1976), The Triassic rocks, with particular reference to predicted and observed performance of some major foundations. Géotechnique 26, 3, p. 391—452.

Meigh, A. C., Greenland, S. W. (1965), In-situ testing of soft rocks. Proc. 6th ICOSMFE, Montreal, 1, p. 73—76.

Melan, E. (1918), Die Verteilung des Druckes durch eine elastische Schicht. Öst. Wochenschrift f. d. öffentl. Baudienst 24, 34, p. 394—397.

Melan, E. (1932), Spannungszustand der durch eine Einzelkraft im Innern beanspruchten Halbscheibe. Zeitschrift f. angew. Math. u. Mech. 12, 6, p. 343—346.

Ménard, L. (1971), Le tassement des fondations et les techniques pressiométriques. Ann. de l'ITBTP, No. 288, p. 105—121.

Ménard, L., Rousseau, J. (1962), L'évaluation des tassements — tendances nouvelles. Sols — Soils, 1, p. 13—30.

Mencl, V. (1955), Soil Mechanics (in Czech). Nakl. ČSAV, Praha, 255 pp.

Mencl, V. (1966), Soil and Rock Mechanics (in Czech). Nakl. ČSAV, Praha, 329 pp.
Meyerhof, G. G. (1965), Shallow foundations. J. SM Div. ASCE, 91, 2, p. 21—31.
Michell, J. H. (1900), The stress distribution in an aeolotropic solid with an infinite plane boundary. Proc. London Math. Soc. 32, p. 247—258.
Milovic, D. M. (1972), Stresses and displacements in an anisotropic layer due to a rigid circular foundation. Géotechnique 22, 1, p. 169—174.
Milovic, D. M., Tournier, J. P. (1971), Stresses and displacements due to rectangular load on a layer of finite thickness. Soils and Foundations 11, 1, p. 1—27.
Milović, D. M. (1971), Stresses in an aeolotropic soil produced by a rectangular foundation. Proc. 4th Conf. on Soil Mechanics, Budapest, 1, p. 669—676.
Mindlin, R. D. (1936), Force at a point in the interior of a semiinfinite solid. J. Appl. Physics 7, 5, p. 195—202.
Mindlin, R. D., Deresiewicz, H. (1953), Elastic spheres in contact under varying oblique forces. J. Appl. Mech. 20, 3, p. 327—344.
Mitchell, J. K. (1969), Discussion. Proc. 7th ICOSMFE, Mexico, 3, p. 159—169.
Monfore, G. E. (1950), An analysis of the stress distribution in and near stress gauges embedded in elastic solids. U.S. Bureau of Reclamation, Report SP 26, June.
Moretto, O. (1969), Discussion. Proc. 7th ICOSMFE, Mexico, 3, p. 357—359.
Morgan, J. R., Gerrard, C. M. (1968), Free field measurements of stresses and strains in soils. Proc. 4th Conf., Austr. Road Res. Board 4, 2, p. 1743—1760.
Morgan, J. R., Gerrard, C. M. (1973), Anisotropy and non-linearity in sand properties. Proc. 8th ICOSMFE, Moscow, 1.2, p. 287—292.
Morgenstern, N. R., Eisenstein, Z. (1970), Methods of estimating lateral loads and deformations. Spec. Conf. ASCE on "Lateral Stresses in the Ground and Design of Earth-Retaining Structures", Cornell University, p. 51—102.
Morgenstern, N. R., Tamuly Phukan, A. L. (1968), Stresses and displacements in a homogeneous non-linear foundation. Proc. Int. Symp. on Rock Mech., Madrid, 1, p. 313—320.
Muller, R. A. (1962), Statistical theory of stress distribution in granular soils. (Муллер, Р. А., 1962, К статистической теории распределения напряжений в зернистом основании). Осн., фунд., и мех. грунтов, 5, p. 4—6.
Myslivec, A., Eichler, J., Jesenák, J. (1970), Soil Mechanics (in Czech). SNTL + Alfa, 388 pp.
Naborczyk, J., Pałka, J. (1975), Active zone in slab foundations substratum. Proc. First Baltic COSMFE, Gdańsk, 2, p. 561—571.
Newmark, N. M. (1942), Influence charts for computation of stresses in elastic soils. Bull. Univ. of Illinois, Eng. Exp. Station, No. 338.
Nishida, Y. (1966), Vertical stress and vertical deformations of ground under a deep circular uniform pressure in the semiinfinite. Proc. 1st Cong. ISRM, Lisbon, 2, p. 493—498.
Novotný, B. (1976), Some aspects of numerical analysis of multilayered halfspace. Acta technica ČSAV 20, 4, p. 382—396.
Obert, L., Merril, R. H., Morgan, T. A. (1962), Borehole deformation gage for determining the stress in mine rock. USBM Rep. Inv. 5978.
Ohde, J. (1939), Zur Theorie der Druckverteilung im Baugrund. Der Bauingenieur 20, 33—34, p. 451—459.
Paál, T. (1971), Bemerkungen zum Ödometerversuch. Proc. 4th Conf. on Soil Mechanics, Budapest, 1, p. 225—234.
Peattie, K. R., Sparrow, R. W. (1954), The fundamental action of earth pressure cells. J. Mech. Phys. Solids 2, 3, p. 141—155.
Peck, R. B., Bazaraa, A. R. S. S. (1969), Discussion. J. SM Div. ASCE 95, 3, p. 305—309.
Peck, R. B., Hanson, W. E., Thornburn, T. H. (1974), Foundation engineering. J. Wiley & Sons, New York, 2nd ed., 514 pp.

PERLEA, V. D. (1969), Discussion. Proc. Conf. "In Situ Investigations in Soils and Rocks", BGS, London 1970, p. 197—199.

PETRIANIN, V. F. (1970), Study of the plane problem of the state of stress and strain in a box. (Петрянин, В. Ф., 1970, Изучение плоской задачи напряженно-деформированного состояния грунтового основания в лотке). Сб. „Вопросы мех. грунтов и строит. на лесс. основаниях", Грозный, p. 98—107.

PETRIANIN, V. F., SARKISOV, I. N., RIABCHENKO, L. N. (1972), On the effect of the state of three-dimensional stresses in subsoil on the output signal of a gauge. (Петрянин, В. Ф., Саркисов, И. Н., Рябченко, Л. Н., 1972, О влиянии вида пространственного напряженного состояния грунта на выходной сигнал датчика). Осн., фунд. и мех. грунтов, 5, p. 4—5.

POIROT, M. (1971), Discussion. Ann. de l'ITBTP, No. 288, p. 124.

POKROVSKIY, G. I. (1937), Soil physics research. (Покровский, Г. И., 1937, Исследования по физике грунтов). ОНТИ г. 1937.

POKROVSKIY, G. I., FEDOROV, I. S. (1936), Studies of soil pressure and soil deformations by means of a centrifuge. Proc. ICOSMFE, Cambridge, Mass., 1, p. 70

POKROVSKIY, G. I., FEDOROV, S. (1953), Solutions of engineering problems by means of centrifugal modelling. (Покровский, Г. И., Федоров, И. С., 1953, Центробежное моделирование для решения инженерных задач). Госстройиздат, Москва, 196 pp.

POLSHIN, D. E., TOKAR, R. A. (1957), Maximum allowable non-uniform settlement of structures. Proc. 4th ICOSMFE, London, 1, p. 402—405.

POLSHIN, D. E., RUDNITSKI, N. Y., CHIZHIKOV, P. G., YAKOVLEVA, T. G. (1973), Centrifugal model testing of foundation soils of building structures. Proc. 8th ICOSMFE, Moscow, 1.3, p. 203—208.

POULOS, H. G. (1967), Stresses and displacements in an elastic layer underlain by a rough rigid base. Géotechnique 17, 4, p. 378—410.

POULOS, H. G. (1968), The behaviour of a rigid circular plate resting on a finite elastic layer. Civ. Eng. Trans., Inst. of Eng. Aust., vol. CE 10, p. 213—219.

POULOS, H. G., DAVIS, E. H. (1972), Laboratory determination of in situ horizontal stress in soil masses. Géotechnique 22, 1, p. 177—182.

POULOS, H. G., DAVIS, E. H. (1974), Elastic solutions for soil and rock mechanics. J. Wiley & Sons, 411 pp.

PRANGE, B. (1965), Beitrag zum Problem der Spannungsmessung im Halbraum. Veröffentl. d. Inst. f. Bodenm. u. Grundbau d. TH Fridericiana in Karlsruhe, No. 18, 159 pp.

PRANGE, B. (1971), The state of telemetry in soil mechanics. Proc. Roscoe Mem. Symp. "Stress-Strain Behaviour of Soils", Cambridge, p. 476—488.

PRESS, H. (1930), Baugrundbelastungsversuche mit Flächen verschiedener Grösse. Die Bautechnik 8, 42, p. 641—643.

PRÉVOST, J.-H., HOËG, K. (1975), Analysis of pressuremeter in strain-softening soil. J. Geot. Eng. Div. 101, 8, p. 717—732.

Price, N. J. (1969), Laws of rock behaviour in the earth's crust. Proc. 11th Symp. Rock Mech., Berkeley, p. 3—23.

PROCTER, D. C., BARTON, R. R. (1974), Measurement of the angle of inter-particle friction. Géotechnique 24, 4, p. 581—604.

RADHAKRISHNAN, N., REESE, L. C. (1970), A review of applications of the finite element method of analysis to problems in soil and rock mechanics. Soils and Foundations 10, 3, p. 95—112.

RAJU, V. S. (1967), Ruhedruck und eindimensionale Zusammendrückung von Kugelpackungen. Veröffentl. d. Inst. f. Bodenm. u. Felsm. d. TH Fridericiana in Karlsruhe, No. 29, 112 pp.

RANKINE, W. J. M. (1857), On the stability of loose earth. Trans. Roy. Soc. London 147, 1, p. 9—27.

RAPPOPORT, R. M. (1948), Boussinesq's problem for a layered elastic halfspace. (Раппопорт, Р. М.,

1948, Задача Буссинеска для слоистого упругого полупространства). Труды Ленинградского политехнического института № 5.

RAYMOND, G. P., TOWNSEND, D. L., LOJKASEK, M. J. (1971), The effect of sampling on the undrained soil properties of a Leda soil. Can. Geot. J. 8, 4, p. 546—557.

RESENDIZ, D., HERRERA, I. (1969), A probabilistic formulation of settlement-controlled design. Proc. 7th ICOSMFE, Mexico, 2, p. 217—225.

ROGATIN, J. A. (1976), Investigation of the mechanical properties of clayey subsoil at different depths. (Рогатин, Ю. А., 1976, К исследованию механических свойств глинистого грунта на различной глубине). Осн., фунд. и мех. грунтов, 5, p. 40—41.

ROSCOE, K. H. (1969), Some soil mechanics concepts and the possibility of their wider application. In: M. Te'eni (Ed.), 1971. Structure, solid mechanics and engineering design. Proc. South. 1969 Civ. Eng. Mat. Conf., Wiley-Interscience, Pt. 1, 81, p. 1—27.

ROSCOE, K. H., BURLAND, J. B. (1968), On the generalized stress-strain behaviour of "wet" clay. In: J. Heyman, F. A. Leckie (Ed.), 1968. Engineering plasticity, Cambridge Univ. Press, p. 535—609.

RYBÁŘ, J. (1971), Tektonisch beeinflusste Hangdeformationen im Braunkohlenbecken. Rock Mech. 3, 3, p. 139—158.

SANGLERAT, G. (1972), The penetrometer and soil exploration. Elsevier, 464 pp.

SANII, H. (1973), Etude de tassement de fondations reposant sur des sols pulvérulents. Ann. de TPB 1, p. 1—40.

SAVVATEIEV, S. S. (1972), Systematic errors of laboratory investigation of deformability of subsident soils. (Савватеев, С. С., 1972, Систематические погрешности лабораторных исследований деформативности просадочных грунтов). Осн., фунд. и мех. грунтов, 3, p. 8—10.

SCHERMAN, K. A. (1969), Discussion. Proc. Conf. "In Situ Investigations in Soils and Rocks", BGS, London 1970, p. 50—51.

SCHIFFMAN, R. L., WHITMAN, R. V., JORDAN, J. C. (1970), Settlement problem oriented computer language. J. SM Div. ASCE 96, 2, p. 649—669.

SCHLEICHER, P. (1926), Zur Theorie des Baugrundes. Der Bauingenieur 7, p. 931 and 949; cf. Kreisplatten auf elastischer Unterlage, J. Springer, Berlin, 1926.

SCHLOSSER, F. (1974), Influence des déformations des grains dans les milieux granulaires. Bull. liaison LPC 69, p. 97—103.

SCHMERTMANN, J. H. (1970), Static cone to compute static settlement over sand. J. SM Div. ASCE 96, 3, p. 1011—1043.

SCHOFIELD, A., WROTH, P. (1968), Critical state soil mechanics. McGraw-Hill, 310 pp.

SCHULTZE, E. (1963), Beispiele für Setzungsbeobachtungen in bindigen und nichtbindigen Böden. Proc. ECOSMFE "Problems of Settlements and Compressibility of Soils", Wiesbaden, 1, p. 143—162.

SCHULTZE, E. (1975), Fundamentals for calculating shallow foundations on sand and silt, especially for hydraulic engineering and harbour constructions. Proc. First Baltic COSMFE, Gdańsk, 1, p. 165—179.

SCHULTZE, E., SHERIF, G. (1973), Prediction of settlements from evaluated settlement observations for sand. Proc. 8th ICOSMFE, Moscow, 1.3, p. 225—230.

SCOTT, R. F. (1963), Principles of soil mechanics. Addison-Wesley, Reading (Mass.), 550 pp.

SEED, H. B. (1965), Settlement analyses, a review of theory and testing procedures. J. SM Div. ASCE 91, 2, p. 39—48.

SHERIF, G. (1973), Setzungsmessungen an Industrie- und Hochbauten und ihre Auswertung. Mitt. Inst. VGB TH Aachen, Heft 57, 251 pp.

SHIROKOV, V. N., SOLOMIN, V. I., CHEREMNIK, V. A., MALYSHEV, M. V., ZARETSKY, YU. K. (1971), A circular rigid plate on a nonlinearly deforming base. Proc. 4th COSMFE, Budapest, p. 757—764.

Shvets, V. B. (1964), On the reliability of the determination of the deformation modulus in clayey soils, established by compression tests. (Швец, В. Б., 1964, О достоверности определения модуля деформации элювиальных глинистых грунтов по данным компрессионных испытаний). Осн., фунд. и мех. грунтов, 5, p. 12—13.

Shvets, V. B., Kul'chitskiy, G. B. (1971), Pecularities of formation of a stress-deformation zone in residual soils of various degrees of weathering. (Швец, В. Б., Кульчицкий, Г. Б., 1971, Особенности формирования напряженно-деформируемой зоны в элювиальных грунтах различной степени выветренности). Сб. конф. „Основания, фундаменты и механика грунтов", Киев, p. 185—188.

Shvetsov, G. I. (1967), A comparison of actual and calculated deformations of the soil surface. (Швецов, Г. И., 1967, Сравнение фактических деформаций поверхности грунтовых оснований с расчетными). Осн., фунд. и мех. грунтов, 3, p. 3—4.

Silveira, A., Souto Silveira, E. B. (1973), Elastic parameters for soils with cross-anisotropy. Proc. 8th ICOSMFE, Moscow, 1.2, p. 361—365.

Simons, N. (1963), Settlement studies on a nine storey apartment building at Økerbråten, Oslo. Proc. ECOSMFE "Problems of Settlements and Compressibility of Soils", Wiesbaden, 1, p. 179—191.

Simons, N. E. (1969), Discussion — session C. Proc. Conf. "In Situ Investigations in Soils and Rocks", BGS, London 1970, p. 200—201.

Simons, N. E. (1974), Normally consolidated and lightly over-consolidated cohesive materials. Review paper — session II. Proc. Conf. "Settlement of Structures", BGS, London 1975, p. 500—530.

Simons, N. E., Rodrigues, J. S. N. (1975), Finite element analysis of the surface deformation due to a uniform loading on a layer of Gibson soil resting on a smooth rigid base. Géotechnique 25, 2, p. 375—379.

Simons, N. E., Som, N. N. (1969), The influence of lateral stresses on the stress-deformation characteristics of London clay. Proc. 7th ICOSMFE, Mexico, 1: 369—377.

Skempton, A. W. (1944), Notes on the compressibility of clays. Quart. J. Geol. Soc. London, vol. C, p. 119—135.

Skempton, A. W. (1953), Soil mechanics in relation to geology. Proc. Yorkshire Geol. Soc. 29, 1, 3, p. 33—62.

Skempton, A. W. (1954), The pore-pressure coefficient A and B. Géotechnique 4, p. 143—147.

Skempton, A. W. (1960), Correspondence on the pore pressure coefficient in saturated soils. Géotechnique 10, 4, p. 186—187.

Skempton, A. W. (1961), Horizontal stress in overconsolidated eocene clay. Proc. 5th ICOSMFE, Paris, 1, p. 351—357.

Skempton, A. W., Bjerrum, L. (1957), A contribution to the settlement analysis of foundations on clay. Géotechnique 7, 4, p. 168—178.

Skempton, A. W., Henkel, D. J. (1957), Tests on London clay from deep borings at Paddington, Victoria and the South Bank. Proc. 4th ICOSMFE, London, 1, p. 100—106.

Skempton, A. W., Henkel, D. J. (1961), Field observation on pore pressures in London clay. Proc. Conf. "Pore Pressure and Suction in Soils", London, p. 81—84.

Skempton, A. W., Peck, R. B., McDonald, D. H. (1965), Settlement analysis of six structures in Chicago and London. Proc. I.C.E. London, 1, 5, 2, p. 166—172.

Smoltczyk, H. U. (1967), Stress computation in soil media. J. SM Div. ASCE 93, 2, p. 101—124.

Smoltczyk, U. (1972), Bodenmechanik zwischen Naturwissenschaft und Technik. Der Bauingenieur 47, 3, p. 73—80.

Sovinc, I. (1961), Stresses and displacements in a limited layer of uniform thickness, resting on a rigid base and subjected to a uniformly distributed flexible load of rectangular shape. Proc. 5th ICOSMFE, Paris, 1, p. 823—827.

Sovinc, I. (1969), Displacements and inclinations of rigid footing resting on a limited elastic layer of uniform thickness. Proc. 7th ICOSMFE, Mexico, 1, p. 385—389.

Sowers, G. F. (1962), Shallow foundations. In: G. A. Leonards (Ed.), Foundation Engineering. McGraw-Hill, New York (1136 pp.), p. 525—632.

Stefanoff, G., Zlatarev, K., Grantcharev, M., Milev, G. (1965), Anticipated and observed settlement of a high chimney. Proc. 6th ICOSMFE, Montreal, 2, p. 202—204.

Steinbrenner, W. (1934), Tafeln zur Setzungsberechnung. Strasse 1, p. 121—124.

Strohschneider, O. (1912), Elastische Druckverteilung und Drucküberschreitung in Schüttungen. Sitzungsberichte der Kaiserlichen Akademie der Wissenschaften. Math.-Naturwiss. Klasse, sect. IIa, vol. 121, 10, p. 299—336.

Sutherland, L. (1971), Le tassement des fondations et les techniques pressiométriques. Ann. de L'ITBTP, No. 288, p. 105—121.

Škopek, J. (1968), Uniformly distributed area load in the interior of a semi-infinite solid. Acta technica ČSAV, 5, p. 459—476.

Šuklje, L. (1969), Rheological aspects of soil mechanics. J. Wiley & Sons, New York, 571 pp.

Taylor, D. W. (1947), Review of pressure distribution theories, earth pressure cell investigations and pressure distribution data. Progress Report, U.S. Waterways Experimental Station.

Taylor, D. W. (1948), Fundamentals of soil mechanics. J. Wiley & Sons, New York, 700 pp.

Thurairajah, A., Roscoe, K. H. (1965), The correlation of triaxial compression test data on cohesionless granular media. Proc. 6th ICOSMFE, Montreal, 1, p. 377—381.

Thurston, C. W., Deresiewicz, H. (1959), Analysis of a compression test of a model of a granular medium. J. Appl. Mech. 26, 2, p. 251—258.

Terzaghi, K. (1925), Erdbaumechanik auf bodenphysikalischer Grundlage. F. Deuticke, Wien, 399 pp.

Terzaghi, K. (1943), Theoretical soil mechanics. J. Wiley & Sons, N.Y, 510 pp.

Terzaghi, K. (1961), Discussion of Prof. A. W. Skempton's paper "Horizontal stresses in an over-consolidated eocene clay". Proc. 5th ICOSMFE, Paris, 3, p. 144—145.

Terzaghi, K., Peck, R. B. (1967), Soil mechanics in engineering practice. J. Wiley & Sons, 2nd ed., 729 pp.

Thomas, H. S. H., Ward, W. H. (1969), The design, construction and performance of a vibrating-wire earth pressure cell. Géotechnique 19, 1, p. 39—51.

Thompson, W. Sir (Lord Kelvin) (1848), Note on the integration of the equations of equilibrium of an elastic solid. Cambridge and Dublin Math. J.; reprinted in: Math. and Phys. Papers, 1, p. 97, 1882.

Timoshenko, S. (1934), Theory of elasticity. Mc-Graw Hill, New York.

Torstensson, B. - A. (1973), The behaviour of a cohesion pile group in soft clay. Proc 8th ICOSMFE, Moscow, 2.1: 237—242.

Tortorici, M., Umiltà, G. (1969), Il calcolo automatico delle tensioni in un terreno di fondazione sollecitato su strisce parallele del piano limite. Rivista Italiana di Geotecnica 3, 2, p. 115—124.

Tschebotarioff, G. P. (1951), Soil mechanics, foundations and earth structures. McGraw-Hill, New York, 655 pp.

Tschebotarioff, G. P. (1973), Discussion — specialty session 5. Proc. 8th ICOSMFE, Moscow, 4.3, p. 227—230.

Tsitovitch, N. A., Baranov, D. S. (1961), On the accuracy of the method for direct pressure measurements in soils. Proc. 5th ICOSMFE, Paris, 3, p. 337—338.

Tsytovich, N. A. (1969), Discussion. Proc. 7th ICOSMFE, Mexico, 3, p. 189—191.

Tsytovich, N. A. (1963), Soil mechanics. (Цытович, Н. А., 1963, Механика грунтов). Госстройиздат, Москва, 636 pp.

Tugaenko, J. F., Khutorianskiy, V. A. (1971), Some results of field investigations of deformations in multi-layered foundation soils. (Тугаенко, Ю. Ф., Хуторянский, В. А., 1971, Некоторые результаты полевых исследований деформаций в многослойных основаниях фундаментов). Сб. конф. „Основания, фундаменты и механика грунтов", Киев, p. 64—69.

TURNBULL, W. J., MAXWELL, A. A., AHLVIN, R. G. (1961), Stresses and deflections in homogeneous soil masses. Proc. 5th ICOSMFE, Paris, 2, p. 337—345.

UESHITA, K., MEYERHOF, G. (1968), Surface displacement of an elastic layer under uniformly distributed loads. Highway Res. Record No. 228, p. 1—10.

UESHITA, K., MATSUI, K., NAGASE, S. (1974), Displacement of overconsolidated clay/sand soil during construction of a building. Proc. Conf. "Settlement of Structures", BGS, London 1975, p. 234—241.

URIEL, A. O. (1973), Discussion. Géotechnique 23, 2, p. 294—298.

U.S. Waterways Experimental Station (1944), Soil pressure cell investigation. Tech. Mem. No. 210-1.

VAID, Y. P., CAMPANELLA, G. (1974), Triaxial and plane strain behaviour of natural clay. J. Geot. Eng. Div. 100, 3, p. 207—224.

VAN WAMBEKE, A. (1975), Les corrélations entre caractéristiques géotechniques. Ann. des TPB, 4, p. 229—244.

VARGA, L. (1965), The compressibility of loess soils. Proc. 6th ICOSMFE, Montreal, 1, p. 395—397.

VARGAS, M. (1961), Foundations of tall buildings on sand in São Paulo (Brazil). Proc. 5th ICOSMFE, Paris, 1, p. 841—843.

VINOKUROV, E. F., MIKULICH, V. A. (1975), Investigation of the state of stress and strain in a deep strip foundation by the finite element method. (Винокуров, Е. Ф., Микулич, В. А., 1975, Исследование напряженно-диформированного состояния заглубленного ленточного фундамента методом конечных элементов). Осн., фунд. и мех. грунтов, 5, p. 34—37.

VOIGHT, B. (1966), Interpretation of in situ measurements. Proc. First Congress ISRM, Lisbon, 3, p. 332—348.

WAKELING, T. R. M. (1974), Discussion. Proc. Conf. "Settlement of Structures", BGS, London 1975, p. 668—669.

WARD, W. H., MARSLAND, A., SAMUELS, S. G. (1965), Properties of the London clay at the Ashford Common shaft. Géotechnique 15, 4, p. 321—344.

WARD, W. H., SAMUELS, S. G., BUTLER, M. E. (1959), Further studies of the properties of London clay. Géotechnique 9, 2, p. 33—58.

WESTERGAARD, H. M. (1938), A problem of elasticity suggested by a problem in soil mechanics: soft material reinforced by numerous horizontal sheets. Contr. Mech. Solids — S. Timoshenko 60th Anniversary Volume. Macmillan, New York.

WHIFFIN, A. C., MORRIS, S. A. H. (1962), Piezoelectric gauge for measuring dynamic stresses under loads. The Engineer, London, 203, p. 741—746.

WIBEL, A. R. (1971), Spannungsverteilung unter flachgegründeten Kreisfundamenten auf trockenem Sand. Veröffentl. d. Inst. f. Bodenm. u. Felsm. d. Universität Karlsruhe, Heft 46, 81 + VIII pp.

WIBEL, A. R., BLINDE, A. (1971), Zur Sohldruckverteilung unter starren quadratischen Fundamenten. Veröffentl. d. Inst. f. Bodenm. u. Felsm. d. Universität Karlsruhe, Heft 48, p. 1—55.

WILKINSON, W. B., BARDEN, L., ROCKE, G. (1969), An assessment of in situ and laboratory tests in predicting the pore pressure in an earth dam. Proc. Conf. "In Situ Investigations in Soils and Rocks", BGS, London 1970, p. 277—284.

WINKLER, E. (1867), Die Lehre von Elastizität und Festigkeit, Praha, 182 pp.

WOLF, K. (1935), Ausbreitung der Kraft in der Halbebene und im Halbraum bei anisotropem Material. ZAMM 15, 5, p. 249—254.

WROTH, C. P. (1971), Some aspects of the elastic behaviour of overconsolidated clay. Proc. Roscoe Mem. Symp. "Stress-Strain Behaviour of Soils", Cambridge, p. 347—361.

Wroth, C. P. (1972), General theories of earth pressures and deformations. Proc. 5th ECOSMFE, Madrid, 2, p. 33—52.

Wroth, C. P. (1977), The predicted performance of soft clay under a trial embankment loading based on the Cam-clay model. In: Gudehus, G. (Ed.), Finite Elements in Geomechanics. J. Wiley & Sons Ltd., p. 191—208. (preprint 1974, 17 pp +9 Figs.)

Wroth, C. P., Hughes, J. M. O. (1973), An instrument for the in-situ measurement of the properties of soft clays. Proc. 8th ICOSMFE, Moscow, 1.2, p. 487—494.

Yamaguchi, H., Kimura, T., Konno, H. (1968), On the contact pressure distribution between rigid loads and an elastic solid underlain by a rigid boundary. Soils and Foundations 8, 3, p. 46—62.

Yegorov, K. E., Nichiporovich, A. A. (1961), Research on the deflexion of foundations. Proc. 5th ICOSMFE, Paris, 1, p. 861—866.

Yudhbir, Varadarajan, A. (1975), Stress-path dependence of deformation moduli of clay. J. Geot. Eng. Div. 101, 3, p. 315—327.

Zaretskiy, J. K. (1967), Theory of soil consolidation. (Зарецкий, Ю. К., 1967, Теория консолидации грунтов). Изд. Наука, Москва, 270 pp.

Zelst, van T. W. (1948), An investigation of the factors affecting laboratory consolidation of clay. Proc. 2nd ICOSMFE, Rotterdam, 7, p. 52—61.

Zienkiewicz, O. C. (1970), The finite element method: from intuition to generality. Appl. Mech. Rev., 23, 3, p. 249—256.

Zienkiewicz, O. C., Cheung, Y. K. (1967), The finite element method in structural and continuum mechanics. McGraw-Hill, New York, 274 pp.

Zienkiewicz, O. C., Naylor, D. J. (1971), The adaptation of critical state soil mechanics theory for use in finite elements. Proc. Roscoe Mem. Symp. "Stress-Strain Behaviour of Soils", Cambridge, p. 537—547.

Zolotarev, G. S., Shariy, A. A. (1973), Stress measurement in clayey soils of slcpe cuttings by the method of partial unloading. (Золотарев, Г. С., Шарий, А. А., 1973, Измерение напряжений в глинистых породах склонов и откосов методом частичной разгрузки). Сб. „Вопросы инженерной геологии и грунтоведения", вып. 3, МГУ Москва, p. 351—356.

Abbreviations used in BIBLIOGRAPHY:

ICOSMFE, ECOSMFE	—	International, European Conference on Soil Mechanics and Foundation Engineering.
COSMFE	—	Conference on Soil Mechanics and Foundation Engineering.
J. SM Div. ASCE	—	Journal of the Soil Mechanics and Foundations Division, Proceedings of the American Society of Civil Engineers.

TABLES FOR THE CALCULATION OF THE STRESS COEFFICIENTS

These tables are an excerpt from a book by J.-P. Giroud "Tables pour le calcul des fondations", Dunod éditeur, Paris, 1972.

Table I. Coefficient K_0

Depth: z/L for the corner, $2z/L$ for the centre	B/L = 0	0.1	0.2	1/3	0.4	0.5	2/3	1	1.5	2	2.5	3	5	10	∞
0	0.000	0.250	0.250	0.250	0.250	0.250	0.250	0.250	0.250	0.250	0.250	0.250	0.250	0.250	0.250
0.2	0.000	0.137	0.204	0.234	0.240	0.244	0.247	0.249	0.249	0.249	0.249	0.249	0.249	0.249	0.249
0.4	0.000	0.076	0.136	0.187	0.202	0.218	0.231	0.240	0.243	0.244	0.244	0.244	0.244	0.244	0.244
0.5	0.000	0.061	0.113	0.164	0.181	0.200	0.218	0.232	0.238	0.239	0.240	0.240	0.240	0.240	0.240
0.6	0.000	0.051	0.096	0.143	0.161	0.182	0.204	0.223	0.231	0.233	0.234	0.234	0.234	0.234	0.234
0.8	0.000	0.037	0.071	0.111	0.127	0.148	0.173	0.200	0.214	0.218	0.219	0.220	0.220	0.220	0.220
1	0.000	0.028	0.055	0.087	0.101	0.120	0.145	0.175	0.194	0.200	0.202	0.203	0.204	0.205	0.205
1.2	0.000	0.022	0.043	0.069	0.081	0.098	0.121	0.152	0.173	0.182	0.185	0.187	0.189	0.189	0.189
1.4	0.000	0.018	0.035	0.056	0.066	0.080	0.101	0.131	0.154	0.164	0.169	0.171	0.174	0.174	0.174
1.5	0.000	0.016	0.031	0.051	0.060	0.073	0.092	0.121	0.145	0.156	0.161	0.164	0.166	0.167	0.167
1.6	0.000	0.014	0.028	0.046	0.055	0.067	0.085	0.112	0.136	0.148	0.154	0.157	0.160	0.160	0.160
1.8	0.000	0.012	0.024	0.039	0.046	0.056	0.072	0.097	0.121	0.133	0.140	0.143	0.147	0.148	0.148
2	0.000	0.010	0.020	0.033	0.039	0.048	0.061	0.084	0.107	0.120	0.127	0.131	0.136	0.137	0.137
2.5	0.000	0.007	0.013	0.022	0.027	0.033	0.043	0.060	0.080	0.093	0.101	0.106	0.113	0.115	0.115
3	0.000	0.005	0.010	0.016	0.019	0.024	0.031	0.045	0.061	0.073	0.081	0.087	0.096	0.099	0.099
4	.0000	0.003	0.006	0.009	0.011	0.014	0.019	0.027	0.038	0.048	0.055	0.060	0.071	0.076	0.076
5	0.000	0.002	0.004	0.006	0.007	0.009	0.012	0.018	0.026	0.033	0.039	0.043	0.055	0.061	0.062
10	0.000	0.000	0.001	0.002	0.002	0.002	0.003	0.005	0.007	0.009	0.011	0.013	0.020	0.028	0.032
15	0.000	0.000	0.000	0.001	0.001	0.001	0.001	0.002	0.003	0.004	0.005	0.006	0.010	0.016	0.021
20	0.000	0.000	0.000	0.000	0.000	0.001	0.001	0.001	0.002	0.002	0.003	0.004	0.006	0.010	0.016
50	0.000	0.000	0.000	0.000	0.000	0.000	0.000	0.000	0.000	0.000	0.000	0.001	0.001	0.002	0.006

(After Giroud)

Table II. Coefficient K_1

Depth: z/L for the corner, $2z/L$ for the centre	B/L = 0	0.1	0.2	1/3	0.4	0.5	2/3	1	1.5	2	2.5	3	5	10	∞
0	0.000	0.159	0.159	0.159	0.159	0.159	0.159	0.159	0.159	0.159	0.159	0.159	0.159	0.159	0.159
0.2	0.000	0.071	0.111	0.135	0.140	0.145	0.149	0.152	0.153	0.153	0.153	0.153	0.153	0.153	0.153
0.4	0.000	0.037	0.067	0.095	0.105	0.115	0.125	0.133	0.136	0.137	0.137	0.137	0.137	0.137	0.137
0.5	0.000	0.028	0.054	0.079	0.089	0.100	0.111	0.121	0.125	0.127	0.127	0.127	0.127	0.127	0.127
0.6	0.000	0.023	0.043	0.066	0.075	0.085	0.097	0.109	0.115	0.116	0.117	0.117	0.117	0.117	0.117
0.8	0.000	0.015	0.029	0.046	0.053	0.062	0.073	0.086	0.093	0.095	0.096	0.097	0.097	0.097	0.097
1	0.000	0.010	0.020	0.032	0.037	0.045	0.054	0.067	0.075	0.077	0.079	0.079	0.079	0.080	0.080
1.2	0.000	0.007	0.014	0.023	0.027	0.033	0.040	0.051	0.059	0.062	0.064	0.064	0.065	0.065	0.065
1.4	0.000	0.005	0.010	0.017	0.020	0.024	0.030	0.040	0.047	0.050	0.052	0.053	0.054	0.054	0.054
1.5	0.000	0.004	0.009	0.014	0.017	0.021	0.026	0.035	0.042	0.045	0.047	0.048	0.049	0.049	0.049
1.6	0.000	0.004	0.008	0.013	0.015	0.018	0.023	0.031	0.038	0.041	0.043	0.044	0.045	0.045	0.045
1.8	0.000	0.003	0.006	0.010	0.011	0.014	0.018	0.024	0.030	0.034	0.035	0.036	0.037	0.038	0.038
2	0.000	0.002	0.004	0.007	0.009	0.011	0.014	0.019	0.025	0.028	0.029	0.030	0.032	0.032	0.032
2.5	0.000	0.001	0.003	0.004	0.005	0.006	0.008	0.011	0.015	0.018	0.019	0.020	0.022	0.022	0.022
3	0.000	0.001	0.002	0.003	0.003	0.004	0.005	0.007	0.010	0.012	0.013	0.014	0.015	0.016	0.016
4	0.000	0.000	0.001	0.001	0.001	0.002	0.002	0.003	0.005	0.006	0.007	0.007	0.009	0.009	0.009
5	0.000	0.000	0.000	0.001	0.001	0.001	0.001	0.002	0.003	0.003	0.004	0.004	0.005	0.006	0.006
10	0.000	0.000	0.000	0.000	0.000	0.000	0.000	0.000	0.000	0.000	0.001	0.001	0.001	0.001	0.002
15	0.000	0.000	0.000	0.000	0.000	0.000	0.000	0.000	0.000	0.000	0.000	0.000	0.000	0.001	0.001
20	0.000	0.000	0.000	0.000	0.000	0.000	0.000	0.000	0.000	0.000	0.000	0.000	0.000	0.000	0.000
50	0.000	0.000	0.000	0.000	0.000	0.000	0.000	0.000	0.000	0.000	0.000	0.000	0.000	0.000	0.000

(After Giroud)

Table III. Coefficient K_2

Depth: z/L for the corner, $2z/L$ for the centre	B/L: 0	0.1	0.2	1/3	0.4	0.5	2/3	1	1.5	2	2.5	3	5	10	∞
0	0.000	0.250	0.250	0.250	0.250	0.250	0.250	0.250	0.250	0.250	0.250	0.250	0.250	0.250	0.250
0.2	0.000	0.069	0.116	0.149	0.159	0.169	0.177	0.184	0.187	0.188	0.188	0.188	0.188	0.188	0.188
0.4	0.000	0.031	0.058	0.085	0.095	0.106	0.118	0.128	0.133	0.134	0.134	0.134	0.135	0.135	0.135
0.5	0.000	0.022	0.043	0.064	0.073	0.083	0.094	0.105	0.110	0.112	0.112	0.112	0.113	0.113	0.113
0.6	0.000	0.017	0.032	0.049	0.056	0.065	0.075	0.086	0.091	0.093	0.093	0.094	0.094	0.094	0.094
0.8	0.000	0.009	0.018	0.029	0.034	0.040	0.047	0.057	0.062	0.064	0.064	0.065	0.065	0.065	0.065
1	0.000	0.006	0.011	0.018	0.021	0.025	0.030	0.037	0.042	0.044	0.045	0.045	0.045	0.045	0.045
1.2	0.000	0.003	0.007	0.011	0.013	0.016	0.020	0.025	0.029	0.031	0.032	0.032	0.032	0.032	0.032
1.4	0.000	0.002	0.004	0.007	0.008	0.010	0.013	0.017	0.020	0.022	0.023	0.023	0.023	0.023	0.023
1.5	0.000	0.002	0.004	0.006	0.007	0.008	0.011	0.014	0.017	0.019	0.020	0.020	0.020	0.020	0.020
1.6	0.000	0.001	0.003	0.005	0.006	0.007	0.009	0.012	0.015	0.016	0.017	0.017	0.017	0.017	0.017
1.8	0.000	0.001	0.002	0.003	0.004	0.005	0.006	0.008	0.011	0.012	0.012	0.013	0.013	0.013	0.013
2	0.000	0.001	0.001	0.002	0.003	0.003	0.004	0.006	0.008	0.009	0.009	0.010	0.010	0.010	0.010
2.5	0.000	0.000	0.001	0.001	0.001	0.002	0.002	0.003	0.004	0.005	0.005	0.005	0.006	0.006	0.006
3	0.000	0.000	0.000	0.001	0.001	0.001	0.001	0.002	0.002	0.003	0.003	0.003	0.003	0.003	0.003
4	0.000	0.000	0.000	0.000	0.000	0.000	0.000	0.001	0.001	0.001	0.001	0.001	0.001	0.002	0.002
5	0.000	0.000	0.000	0.000	0.000	0.000	0.000	0.000	0.000	0.000	0.001	0.001	0.001	0.001	0.001
10	0.000	0.000	0.000	0.000	0.000	0.000	0.000	0.000	0.000	0.000	0.000	0.000	0.000	0.000	0.000
15	0.000	0.000	0.000	0.000	0.000	0.000	0.000	0.000	0.000	0.000	0.000	0.000	0.000	0.000	0.000
20	0.000	0.000	0.000	0.000	0.000	0.000	0.000	0.000	0.000	0.000	0.000	0.000	0.000	0.000	0.000
50	0.000	0.000	0.000	0.000	0.000	0.000	0.000	0.000	0.000	0.000	0.000	0.000	0.000	0.000	0.000

(After Giroud)

Table IV. Coefficient K_3

Depth z/L for the corner, $2z/L$ for the centre	B/L: 0	0.1	0.2	1/3	0.4	0.5	2/3	1	1.5	2	2.5	3	5	10	∞
0	0.000	∞	∞	∞	∞	∞	∞	∞	∞	∞	∞	∞	∞	∞	∞
0.2	0.000	0.107	0.189	0.259	0.282	0.307	0.332	0.353	0.362	0.364	0.365	0.365	0.365	0.366	0.366
0.4	0.000	0.037	0.069	0.104	0.177	0.133	0.150	0.167	0.175	0.177	0.177	0.178	0.178	0.178	0.178
0.5	0.000	0.023	0.045	0.069	0.079	0.091	0.104	0.118	0.125	0.127	0.128	0.129	0.129	0.129	0.129
0.6	0.000	0.016	0.030	0.047	0.054	0.063	0.074	0.085	0.091	0.093	0.094	0.094	0.094	0.095	0.095
0.8	0.000	0.007	0.014	0.023	0.026	0.031	0.038	0.045	0.050	0.052	0.052	0.052	0.053	0.053	0.053
1	0.000	0.004	0.007	0.012	0.014	0.016	0.020	0.025	0.028	0.030	0.030	0.030	0.031	0.031	0.031
1.2	0.000	0.002	0.004	0.006	0.007	0.009	0.011	0.014	0.017	0.018	0.018	0.018	0.019	0.019	0.019
1.4	0.000	0.001	0.002	0.004	0.004	0.005	0.006	0.009	0.010	0.011	0.011	0.012	0.012	0.012	0.012
1.5	0.000	0.001	0.002	0.003	0.003	0.004	0.005	0.007	0.008	0.009	0.009	0.009	0.010	0.010	0.010
1.6	0.000	0.001	0.001	0.002	0.003	0.003	0.004	0.005	0.006	0.007	0.007	0.008	0.008	0.008	0.008
1.8	0.000	0.000	0.001	0.001	0.002	0.002	0.002	0.003	0.004	0.005	0.005	0.005	0.005	0.005	0.005
2	0.000	0.000	0.001	0.001	0.001	0.001	0.002	0.002	0.003	0.003	0.003	0.004	0.004	0.004	0.004
5.2	0.000	0.000	0.000	0.000	0.000	0.000	0.001	0.001	0.001	0.001	0.001	0.002	0.002	0.002	0.002
3	0.000	0.000	0.000	0.000	0.000	0.000	0.000	0.000	0.001	0.001	0.001	0.001	0.001	0.001	0.001
4	0.000	0.000	0.000	0.000	0.000	0.000	0.000	0.000	0.000	0.000	0.000	0.000	0.000	0.000	0.000
5	0.000	0.000	0.000	0.000	0.000	0.000	0.000	0.000	0.000	0.000	0.000	0.000	0.000	0.000	0.000
10	0.000	0.000	0.000	0.000	0.000	0.000	0.000	0.000	0.000	0.000	0.000	0.000	0.000	0.000	0.000
15	0.000	0.000	0.000	0.000	0.000	0.000	0.000	0.000	0.000	0.000	0.000	0.000	0.000	0.000	0.000
20	0.000	0.000	0.000	0.000	0.000	0.000	0.000	0.000	0.000	0.000	0.000	0.000	0.000	0.000	0.000
50	0.000	−0.000	−0.000	−0.000	−0.000	−0.000	0.000	−0.000	−0.000	0.000	0.000	−0.000	0.000	0.000	0.000

(After Giroud)

Table V. Coefficient K_5

Depth: z/L for the corner, $2z/L$ for the centre	B/L: 0	0.1	0.2	1/3	0.4	0.5	2/3	1	1.5	2	2.5	3	5	10	∞
0	0.000	∞	∞	∞	∞	∞	∞	∞	∞	∞	∞	∞	∞	∞	∞
0.2	0.000	0.005	0.027	0.066	0.085	0.109	0.142	0.185	0.218	0.234	0.242	0.247	0.255	0.258	0.259
0.4	0.000	0.001	0.005	0.017	0.025	0.038	0.059	0.091	0.118	0.133	0.141	0.145	0.153	0.156	0.158
0.5	0.000	0.000	0.003	0.010	0.015	0.024	0.039	0.065	0.090	0.104	0.111	0.116	0.123	0.127	0.128
0.6	0.000	0.000	0.002	0.006	0.009	0.015	0.026	0.048	0.070	0.082	0.089	0.094	0.101	0.105	0.106
0.8	0.000	0.000	0.001	0.002	0.004	0.007	0.013	0.026	0.042	0.053	0.059	0.063	0.070	0.074	0.075
1	0.000	0.000	0.000	0.001	0.002	0.003	0.007	0.015	0.027	0.035	0.040	0.044	0.051	0.054	0.055
1.2	0.000	0.000	0.000	0.001	0.001	0.002	0.004	0.009	0.017	0.024	0.028	0.032	0.038	0.041	0.042
1.4	0.000	0.000	0.000	0.000	0.001	0.001	0.002	0.005	0.011	0.016	0.020	0.023	0.029	0.032	0.033
1.5	0.000	0.000	0.000	0.000	0.000	0.001	0.002	0.004	0.009	0.013	0.017	0.020	0.025	0.028	0.029
1.6	0.000	0.000	0.000	0.000	0.000	0.001	0.001	0.003	0.007	0.011	0.015	0.017	0.022	0.025	0.026
1.8	0.000	0.000	0.000	0.000	0.000	0.000	0.001	0.002	0.005	0.008	0.011	0.013	0.017	0.020	0.021
2	0.000	0.000	0.000	0.000	0.000	0.000	0.000	0.001	0.003	0.006	0.008	0.010	0.014	0.017	0.018
2.5	0.000	0.000	0.000	0.000	0.000	0.000	0.000	0.001	0.001	0.003	0.004	0.005	0.008	0.011	0.012
3	0.000	0.000	0.000	0.000	0.000	0.000	0.000	0.000	0.001	0.001	0.002	0.003	0.005	0.007	0.008
4	0.000	0.000	0.000	0.000	0.000	0.000	0.000	0.000	0.000	0.000	0.001	0.001	0.002	0.004	0.005
5	0.000	0.000	0.000	0.000	0.000	0.000	0.000	0.000	0.000	0.000	0.000	0.000	0.001	0.002	0.003
10	0.000	−0.000	0.000	0.000	0.000	0.000	0.000	0.000	0.000	0.000	0.000	0.000	0.000	0.000	0.001
15	0.000	−0.000	−0.000	0.000	0.000	0.000	0.000	0.000	0.000	0.000	0.000	0.000	0.000	0.000	0.000
20	0.000	−0.000	−0.000	−0.000	−0.000	0.000	0.000	0.000	0.000	0.000	0.000	0.000	0.000	0.000	0.000
50	0.000	−0.000	−0.000	−0.000	−0.000	−0.000	0.000	−0.000	−0.000	0.000	0.000	0.000	0.000	0.000	0.000

(After Giroud)

Table VI. Coefficient K'_2

Depth: z/L for the corner, $2z/L$ for the centre	B/L: 0	0.1	0.2	1/3	0.4	0.5	2/3	1	1.5	2	2.5	3	5	10	∞
0	0.000	0.234	0.219	0.199	0.189	0.176	0.156	0.125	0.094	0.074	0.061	0.051	0.031	0.016	0.000
0.2	0.000	0.059	0.097	0.118	0.121	0.122	0.118	0.103	0.082	0.067	0.056	0.048	0.030	0.016	0.000
0.4	0.000	0.026	0.048	0.069	0.075	0.082	0.086	0.083	0.071	0.060	0.051	0.045	0.029	0.015	0.000
0.5	0.000	0.019	0.036	0.054	0.060	0.067	0.073	0.074	0.066	0.056	0.049	0.043	0.028	0.015	0.000
0.6	0.000	0.015	0.028	0.043	0.049	0.056	0.062	0.066	0.061	0.053	0.047	0.041	0.028	0.015	0.000
0.8	0.000	0.009	0.018	0.029	0.033	0.039	0.046	0.052	0.052	0.047	0.043	0.038	0.026	0.015	0.000
1	0.000	0.007	0.013	0.021	0.024	0.029	0.035	0.042	0.044	0.042	0.039	0.035	0.025	0.014	0.000
1.2	0.000	0.005	0.009	0.015	0.018	0.022	0.027	0.034	0.037	0.037	0.035	0.032	0.024	0.014	0.000
1.4	0.000	0.004	0.007	0.012	0.014	0.017	0.021	0.027	0.032	0.033	0.032	0.030	0.023	0.014	0.000
1.5	0.000	0.003	0.006	0.010	0.012	0.015	0.019	0.025	0.029	0.031	0.030	0.029	0.022	0.014	0.000
1.6	0.000	0.003	0.006	0.009	0.011	0.013	0.017	0.023	0.027	0.029	0.028	0.027	0.022	0.013	0.000
1.8	0.000	0.002	0.005	0.007	0.009	0.011	0.014	0.019	0.024	0.025	0.026	0.025	0.021	0.013	0.000
2	0.000	0.002	0.004	0.006	0.007	0.009	0.012	0.016	0.020	0.023	0.023	0.023	0.020	0.013	0.000
2.5	0.000	0.001	0.002	0.004	0.005	0.006	0.008	0.011	0.015	0.017	0.018	0.019	0.017	0.012	0.000
3	0.000	0.001	0.002	0.003	0.003	0.004	0.006	0.008	0.011	0.013	0.015	0.015	0.015	0.011	0.000
4	0.000	0.000	0.001	0.002	0.002	0.002	0.003	0.005	0.007	0.008	0.010	0.011	0.012	0.010	0.000
5	0.000	0.000	0.001	0.001	0.001	0.002	0.002	0.003	0.004	0.006	0.007	0.008	0.009	0.009	0.000
10	0.000	0.000	0.000	0.000	0.000	0.000	0.001	0.001	0.001	0.002	0.002	0.002	0.003	0.005	0.000
15	0.000	0.000	0.000	0.000	0.000	0.000	0.000	0.000	0.001	0.001	0.001	0.001	0.002	0.003	0.000
20	0.000	0.000	0.000	0.000	0.000	0.000	0.000	0.000	0.000	0.000	0.000	0.001	0.001	0.002	0.000
50	0.000	0.000	0.000	0.000	0.000	0.000	0.000	0.000	0.000	0.000	0.000	0.000	0.000	0.000	0.000

(After Giroud)

Table VII. Coefficient K_3'

Depth: z/L for the corner, $2z/L$ for the centre	B/L: 0	0.1	0.2	1/3	0.4	0.5	2/3	1	1.5	2	2.5	3	5	10	∞
0	0.000	0.143	0.128	0.109	0.100	0.088	0.071	0.047	0.027	0.017	0.011	0.008	0.003	0.001	0.000
0.2	0.000	0.025	0.040	0.048	0.049	0.048	0.044	0.033	0.021	0.014	0.010	0.007	0.003	0.001	0.000
0.4	0.000	0.009	0.016	0.023	0.025	0.026	0.026	0.023	0.016	0.011	0.008	0.006	0.003	0.001	0.000
0.5	0.000	0.006	0.011	0.016	0.018	0.020	0.021	0.019	0.014	0.010	0.008	0.006	0.003	0.001	0.000
0.6	0.000	0.004	0.008	0.012	0.013	0.015	0.016	0.016	0.012	0.009	0.007	0.005	0.002	0.001	0.000
0.8	0.000	0.002	0.004	0.007	0.008	0.009	0.010	0.011	0.009	0.007	0.006	0.005	0.002	0.001	0.000
1	0.000	0.001	0.003	0.004	0.005	0.006	0.007	0.008	0.007	0.006	0.005	0.004	0.002	0.001	0.000
1.2	0.000	0.001	0.002	0.003	0.003	0.004	0.005	0.006	0.006	0.005	0.004	0.004	0.002	0.001	0.000
1.4	0.000	0.001	0.001	0.002	0.002	0.003	0.003	0.004	0.004	0.004	0.004	0.003	0.002	0.001	0.000
1.5	0.000	0.000	0.001	0.002	0.002	0.002	0.003	0.004	0.004	0.004	0.003	0.003	0.002	0.001	0.000
1.6	0.000	0.000	0.001	0.001	0.002	0.002	0.002	0.003	0.003	0.003	0.003	0.003	0.002	0.001	0.000
1.8	0.000	0.000	0.001	0.001	0.001	0.001	0.002	0.002	0.003	0.003	0.003	0.002	0.001	0.001	0.000
2	0.000	0.000	0.000	0.001	0.001	0.001	0.001	0.002	0.002	0.002	0.002	0.002	0.001	0.001	0.000
2.5	0.000	0.000	0.000	0.000	0.000	0.001	0.001	0.001	0.001	0.001	0.001	0.001	0.001	0.000	0.000
3	0.000	0.000	0.000	0.000	0.000	0.000	0.000	0.001	0.001	0.001	0.001	0.001	0.001	0.000	0.000
4	0.000	0.000	0.000	0.000	0.000	0.000	0.000	0.000	0.000	0.000	0.001	0.001	0.001	0.000	0.000
5	0.000	0.000	0.000	0.000	0.000	0.000	0.000	0.000	0.000	0.000	0.000	0.000	0.000	0.000	0.000
10	0.000	0.000	0.000	0.000	0.000	0.000	0.000	0.000	0.000	0.000	0.000	0.000	0.000	0.000	0.000
15	0.000	0.000	0.000	0.000	0.000	0.000	0.000	0.000	0.000	0.000	0.000	0.000	0.000	0.000	0.000
20	0.000	0.000	0.000	0.000	0.000	0.000	0.000	0.000	0.000	0.000	0.000	0.000	0.000	0.000	0.000
50	0.000	−0.000	0.000	0.000	−0.000	0.000	0.000	0.000	0.000	0.000	0.000	0.000	0.000	0.000	0.000

(After Giroud)

Table VIII. Coefficient K'_5

Depth: z/L for the corner, $2z/L$ for the centre	B/L: 0	0.1	0.2	1/3	0.4	0.5	2/3	1	1.5	2	2.5	3	5	10	∞
0	0.000	∞	∞	∞	∞	∞	∞	∞	∞	∞	∞	∞	∞	∞	∞
0.2	0.000	0.036	0.069	0.105	0.120	0.139	0.164	0.197	0.223	0.236	0.244	0.248	0.255	0.258	0.259
0.4	0.000	0.016	0.031	0.049	0.058	0.069	0.085	0.107	0.127	0.137	0.143	0.147	0.153	0.157	0.158
0.5	0.000	0.011	0.023	0.036	0.043	0.051	0.064	0.083	0.100	0.109	0.115	0.118	0.124	0.127	0.128
0.6	0.000	0.009	0.017	0.028	0.032	0.039	0.050	0.065	0.080	0.088	0.093	0.096	0.102	0.105	0.106
0.8	0.000	0.005	0.010	0.017	0.020	0.024	0.031	0.042	0.053	0.059	0.063	0.066	0.071	0.074	0.075
1	0.000	0.003	0.007	0.011	0.013	0.016	0.020	0.028	0.036	0.041	0.045	0.047	0.052	0.054	0.055
1.2	0.000	0.002	0.004	0.007	0.009	0.011	0.014	0.019	0.025	0.030	0.033	0.035	0.039	0.041	0.042
1.4	0.000	0.002	0.003	0.005	0.006	0.007	0.010	0.014	0.018	0.022	0.024	0.026	0.030	0.032	0.033
1.5	0.000	0.001	0.003	0.004	0.005	0.006	0.008	0.012	0.016	0.019	0.021	0.023	0.026	0.028	0.029
1.6	0.000	0.001	0.002	0.004	0.004	0.005	0.007	0.010	00.14	0.016	0.019	0.020	0.023	0.025	0.026
1.8	0.000	0.001	0.002	0.003	0.003	0.004	0.005	0.008	0.010	0.013	0.014	0.016	0.019	0.020	0.021
2	0.000	0.001	0.001	0.002	0.002	0.003	0.004	0.006	0.008	0.010	0.011	0.012	0.015	0.017	0.018
2.5	0.000	0.000	0.001	0.001	0.001	0.002	0.002	0.003	0.005	0.006	0.007	0.007	0.009	0.011	0.012
3	0.000	0.000	0.000	0.001	0.001	0.001	0.001	0.002	0.003	0.004	0.004	0.005	0.006	0.008	0.008
4	0.000	0.000	0.000	0.000	0.000	0.000	0.001	0.001	0.001	0.002	0.002	0.002	0.003	0.004	0.005
5	0.000	0.000	0.000	0.000	0.000	0.000	0.000	0.000	0.001	0.001	0.001	0.001	0.002	0.003	0.003
10	0.000	0.000	0.000	0.000	0.000	0.000	0.000	0.000	0.000	0.000	0.000	0.000	0.000	0.000	0.001
15	0.000	0.000	0.000	0.000	0.000	0.000	0.000	0.000	0.000	0.000	0.000	0.000	0.000	0.000	0.000
20	0.000	0.000	0.000	0.000	0.000	0.000	0.000	0.000	0.000	0.000	0.000	0.000	0.000	0.000	0.000
50	0.000	0.000	0.000	0.000	0.000	0.000	0.000	0.000	0.000	0.000	0.000	0.000	0.000	0.000	0.000

(After Giroud)

Table IX. Coefficient L_2

Depth z/L for the corner, $2z/L$ for the centre	B/L = 0	0.1	0.2	1/3	0.4	0.5	2/3	1	1.5	2	2.5	3	5	10	∞
0	0.000	0.250	0.250	0.250	0.250	0.250	0.250	0.250	0.250	0.250	0.250	0.250	0.250	0.250	0.250
0.2	0.000	0.010	0.045	0.094	0.112	0.134	0.158	0.184	0.201	0.208	0.211	0.214	0.217	0.218	0.219
0.4	0.000	0.002	0.010	0.032	0.045	0.064	0.091	0.128	0.156	0.169	0.176	0.179	0.186	0.188	0.189
0.5	0.000	0.001	0.006	0.020	0.029	0.044	0.068	0.105	0.136	0.151	0.159	0.164	0.172	0.175	0.176
0.6	0.000	0.000	0.003	0.013	0.019	0.031	0.051	0.086	0.118	0.134	0.144	0.144	0.158	0.163	0.164
0.8	0.000	0.000	0.001	0 006	0.009	0.016	0.029	0.057	0.087	0.106	0.117	0.124	0.135	0.141	0.143
1	0.000	0.000	0.001	0.003	0.005	0.009	0.017	0.037	0.064	0.083	0.095	0.103	0.116	0.123	0.125
1.2	0.000	0.000	0.000	0.002	0.003	0.005	0.011	0.025	0.047	0.065	0.077	0.085	0.100	0.108	0.111
1.4	0.000	0.000	0.000	0.001	0.002	0.003	0.007	0.017	0.035	0.051	0.062	0.071	0.087	0.095	0.099
1.5	0.000	0.000	0.000	0.001	0.001	0.003	0.005	0.014	0.030	0.045	0.056	0.064	0.081	0.090	0.094
1.6	0.000	0.000	0.000	0.001	0.001	0.002	0.004	0.012	0.026	0.040	0.051	0.059	0.076	0.085	0.089
1.8	0.000	0.000	0.000	0.000	0.001	0.001	0.003	0.008	0.020	0.031	0.041	0.049	0.066	0.077	0.081
2	0.000	0.000	0.000	0.000	0.001	0.001	0.002	0.006	0.015	0.025	0.034	0.041	0.058	0.069	0.074
2.5	0.000	0.000	0.000	0.000	0.000	0.000	0.001	0.003	0.008	0.014	0.021	0.027	0.043	0.055	0.061
3	0.000	0.000	0.000	0.000	0.000	0.000	0.000	0.002	0.004	0.008	0.013	0.018	0.032	0.045	0.051
4	0.000	0.000	0.000	0.000	0.000	0.000	0.000	0.001	0.002	0.003	0.006	0.008	0.018	0.031	0.039
5	0.000	0.000	0.000	0.000	0.000	0.000	0.000	0.000	0.001	0.002	0.003	0.004	0.011	0.022	0.031
10	0.000	0.000	0.000	0.000	0.000	0.000	0.000	0.000	0.000	0.000	0.000	0.000	0.001	0.006	0.016
15	0.000	0.000	0.000	0.000	0.000	0.000	0.000	0.000	0.000	0.000	0.000	0.000	0.000	0.002	0.011
20	0.000	0.000	0.000	0.000	0.000	0.000	0.000	0.000	0.000	0.000	0.000	0.000	0.000	0.001	0.008
50	0.000	0.000	0.000	0.000	0.000	0.000	0.000	0.000	0.000	0.000	0.000	0.000	0.000	0.000	0.003

(After Giroud)

Table X. Coefficient L'_2

Depth: z/L for the corner, $2z/L$ for the centre	B/L = 0	0.1	0.2	1/3	0.4	0.5	2/3	1	1.5	2	2.5	3	5	10	∞
0	0.000	0.016	0.031	0.051	0.061	0.074	0.094	0.125	0.156	0.176	0.189	0.199	0.219	0.234	0.250
0.2	0.000	0.013	0.025	0.041	0.049	0.060	0.076	0.103	0.130	0.148	0.160	0.169	0.188	0.203	0.219
0.4	0.000	0.010	0.020	0.032	0.039	0.047	0.061	0.083	0.106	0.122	0.133	0.141	0.159	0.174	0.189
0.5	0.000	0.009	0.017	0.029	0.034	0.042	0.054	0.074	0.096	0.111	0.121	0.129	0.146	0.161	0.176
0.6	0.000	0.008	0.015	0.025	0.030	0.037	0.048	0.066	0.086	0.100	0.110	0.118	0.134	0.149	0.164
0.8	0.000	0.006	0.012	0.020	0.023	0.029	0.037	0.052	0.069	0.082	0.091	0.098	0.114	0.127	0.143
1	0.000	0.005	0.009	0.015	0.018	0.023	0.029	0.042	0.056	0.067	0.075	0.082	0.097	0.110	0.125
1.2	0.000	0.004	0.007	0.012	0.015	0.018	0.024	0.034	0.046	0.056	0.063	0.069	0.083	0.096	0.111
1.4	0.000	0.003	0.006	0.010	0.012	0.015	0.019	0.027	0.038	0.046	0.053	0.058	0.072	0.084	0.099
1.5	0.000	0.003	0.005	0.009	0.011	0.013	0.017	0.025	0.035	0.043	0.049	0.054	0.067	0.079	0.094
1.6	0.000	0.002	0.005	0.008	0.010	0.012	0.016	0.023	0.032	0.039	0.045	0.050	0.062	0.074	0.089
1.8	0.000	0.002	0.004	0.007	0.008	0.010	0.013	0.019	0.027	0.033	0.039	0.043	0.055	0.066	0.081
2	0.000	0.002	0.003	0.006	0.007	0.008	0.011	0.016	0.023	0.029	0.033	0.038	0.048	0.059	0.074
2.5	0.000	0.001	0.002	0.004	0.005	0.006	0.007	0.011	0.016	0.020	0.024	0.027	0.036	0.047	0.061
3	0.000	0.001	0.002	0.003	0.003	0.004	0.005	0.008	0.012	0.015	0.018	0.021	0.028	0.038	0.051
4	0.000	0.000	0.001	0.002	0.002	0.002	0.003	0.005	0.007	0.009	0.011	0.013	0.018	0.026	0.039
5	0.000	0.000	0.001	0.001	0.001	0.002	0.002	0.003	0.005	0.006	0.007	0.009	0.013	0.019	0.031
10	0.000	0.000	0.000	0.000	0.000	0.000	0.001	0.001	0.001	0.002	0.002	0.002	0.004	0.007	0.016
15	0.000	0.000	0.000	0.000	0.000	0.000	0.000	0.000	0.001	0.001	0.001	0.001	0.002	0.003	0.011
20	0.000	0.000	0.000	0.000	0.000	0.000	0.000	0.000	0.000	0.000	0.000	0.001	0.001	0.002	0.008
50	0.000	0.000	0.000	0.000	0.000	0.000	0.000	0.000	0.000	0.000	0.000	0.000	0.000	0.000	0.003

(After Giroud)

Table XI. Coefficient M_0

Depth z/L	B/L														
	0	0.1	0.2	1/3	0.4	0.5	2/3	1	1.5	2	2.5	3	5	10	∞
0	0.000	0.250	0.250	0.250	0.250	0.250	0.250	0.250	0.250	0.250	0.250	0.250	0.250	0.250	0.250
0.2	0.000	0.109	0.160	0.180	0.183	0.186	0.187	0.188	0.188	0.188	0.188	0.188	0.188	0.188	0.188
0.4	0.000	0.047	0.083	0.111	0.118	0.126	0.131	0.134	0.134	0.135	0.135	0.135	0.135	0.135	0.135
0.5	0.000	0.033	0.060	0.085	0.092	0.100	0.107	0.111	0.112	0.113	0.113	0.113	0.113	0.113	0.113
0.6	0.000	0.023	0.044	0.065	0.072	0.079	0.087	0.092	0.093	0.094	0.094	0.094	0.094	0.094	0.094
0.8	0.000	0.013	0.025	0.038	0.043	0.049	0.056	0.062	0.064	0.065	0.065	0.065	0.065	0.065	0.065
1	0.000	0.007	0.015	0.023	0.026	0.031	0.036	0.042	0.045	0.045	0.045	0.045	0.045	0.045	0.045
1.2	0.000	0.005	0.009	0.014	0.017	0.020	0.024	0.029	0.031	0.032	0.032	0.032	0.032	0.032	0.032
1.4	0.000	0.003	0.006	0.009	0.011	0.013	0.016	0.020	0.022	0.023	0.023	0.023	0.023	0.023	0.023
1.5	0.000	0.002	0.005	0.007	0.009	0.011	0.013	0.017	0.019	0.020	0.020	0.020	0.020	0.020	0.020
1.6	0.000	0.002	0.004	0.006	0.007	0.008	0.011	0.014	0.016	0.017	0.017	0.017	0.017	0.017	0.017
1.8	0.000	0.001	0.003	0.004	0.005	0.006	0.008	0.010	0.012	0.013	0.013	0.013	0.013	0.013	0.013
2	0.000	0.001	0.002	0.003	0.003	0.004	0.005	0.007	0.009	0.010	0.010	0.010	0.010	0.010	0.010
2.5	0.000	0.000	0.001	0.001	0.002	0.002	0.003	0.004	0.005	0.005	0.005	0.006	0.006	0.006	0.006
3	0.000	0.000	0.000	0.001	0.001	0.001	0.001	0.002	0.003	0.003	0.003	0.003	0.003	0.003	0.003
4	0.000	0.000	0.000	0.000	0.000	0.000	0.000	0.001	0.001	0.001	0.001	0.001	0.002	0.002	0.002
5	0.000	0.000	0.000	0.000	0.000	0.000	0.000	0.000	0.000	0.001	0.001	0.001	0.001	0.001	0.001
10	0.000	0.000	0.000	0.000	0.000	0.000	0.000	0.000	0.000	0.000	0.000	0.000	0.000	0.000	0.000
15	0.000	0.000	0.000	0.000	0.000	0.000	0.000	0.000	0.000	0.000	0.000	0.000	0.000	0.000	0.000
20	0.000	0.000	0.000	0.000	0.000	0.000	0.000	0.000	0.000	0.000	0.000	0.000	0.000	0.000	0.000
50	0.000	0.000	0.000	0.000	0.000	0.000	0.000	0.000	0.000	0.000	0.000	0.000	0.000	0.000	0.000

(After Giroud)

Table XII. Coefficient M_1

Depth: z/L for the corner, $2z/L$ for the centre	B/L: 0	0.1	0.2	1/3	0.4	0.5	2/3	1	1.5	2	2.5	3	5	10	∞
0	0.000	0.159	0.159	0.159	0.159	0.159	0.159	0.159	0.159	0.159	0.159	0.159	0.159	0.159	0.159
0.2	0.000	0.043	0.065	0.075	0.076	0.078	0.078	0.078	0.078	0.078	0.078	0.078	0.078	0.078	0.078
0.4	0.000	0.012	0.021	0.027	0.029	0.030	0.031	0.030	0.030	0.030	0.030	0.030	0.030	0.030	0.030
0.5	0.000	0.006	0.011	0.015	0.016	0.017	0.017	0.016	0.015	0.015	0.015	0.015	0.015	0.015	0.015
0.6	0.000	0.003	0.005	0.007	0.007	0.008	0.007	0.006	0.005	0.005	0.005	0.005	0.005	0.004	0.004
0.8	0.000	−0.000	−0.000	−0.001	−0.001	−0.002	−0.003	−0.004	−0.006	−0.006	−0.007	−0.007	−0.007	−0.007	−0.007
1	0.000	−0.001	−0.002	−0.003	−0.004	−0.005	−0.006	−0.008	−0.010	−0.011	−0.011	−0.011	−0.011	−0.011	−0.011
1.2	0.000	−0.001	−0.002	−0.004	−0.004	−0.005	−0.007	−0.009	−0.011	−0.011	−0.012	−0.012	−00.12	−0.012	−0.012
1.4	0.000	−0.001	−0.002	−0.003	−0.004	−0.005	−0.006	−0.008	−0.011	−0.011	−0.011	−0.012	−0.012	−0.012	−0.012
1.5	0.000	−0.001	−0.002	−0.003	−0.004	−0.004	−0.006	−0.008	−0.009	−0.010	−0.011	−0.011	−0.011	−0.011	−0.011
1.6	0.000	−0.001	−0.002	−0.003	−0.003	−0.004	−0.005	−0.007	−0.009	−0.010	−0.010	−0.011	−0.011	−0.011	−0.011
1.8	0.000	−0.001	−0.001	−0.002	−0.003	−0.003	−0.004	−0.006	−0.008	−0.009	−0.009	−0.009	−0.010	−0.010	−0.010
2	0.000	−0.001	−0.001	−0.002	−0.002	−0.003	−0.004	−0.005	−0.007	−0.007	−0.008	−0.008	−0.009	−0.009	−0.009
2.5	0.000	−0.000	−0.001	−0.001	−0.001	−0.002	−0.002	−0.003	−0.004	−0.005	−0.006	−0.006	−0.006	−0.006	−0.006
3	0.000	−0.000	−0.000	−0.001	−0.001	−0.001	−0.001	−0.002	−0.003	−0.004	−0.004	−0.004	−0.005	−0.005	−0.005
4	0.000	−0.000	−0.000	−0.000	−0.000	−0.001	−0.001	−0.001	−0.001	−0.002	−0.002	−0.002	−0.003	−0.003	−0.003
5	0.000	−0.000	−0.000	−0.000	−0.000	−0.000	−0.000	−0.001	−0.001	−0.001	−0.001	−0.001	−0.002	−0.002	−0.002
10	0.000	−0.000	−0.000	−0.000	−0.000	−0.000	−0.000	−0.000	−0.000	−0.000	−0.000	−0.000	−0.000	−0.000	−0.001
15	0.000	−0.000	−0.000	−0.000	−0.000	−0.000	−0.000	−0.000	−0.000	−0.000	−0.000	−0.000	−0.000	−0.000	−0.000
20	0.000	−0.000	−0.000	−0.000	−0.000	−0.000	−0.000	−0.000	−0.000	−0.000	−0.000	−0.000	−0.000	−0.000	−0.000
50	0.000	−0.000	−0.000	−0.000	−0.000	−0.000	−0.000	−0.000	−0.000	−0.000	−0.000	−0.000	−0.000	−0.000	−0.000

(After Giroud)

Table XIII. Coefficient M_2

Depth z/L	B/L: 0	0.1	0.2	1/3	0.4	0.5	2/3	1	1.5	2	2.5	3	5	10	∞
0	0.000	0.250	0.250	0.250	0.250	0.250	0.250	0.250	0.250	0.250	0.250	0.250	0.250	0.250	0.250
0.2	0.000	0.026	0.040	0.046	0.046	0.046	0.045	0.043	0.042	0.042	0.042	0.042	0.042	0.042	0.042
0.4	0.000	0.002	0.003	0.002	−0.001	−0.000	−0.002	−0.005	−0.007	−0.008	−0.008	−0.008	−0.008	−0.008	−0.008
0.5	0.000	−0.001	−0.002	−0.005	−0.006	−0.008	−0.010	−0.013	−0.015	−0.016	−0.016	−0.016	−0.016	−0.016	−0.016
0.6	0.000	−0.002	−0.004	−0.007	−0.009	−0.010	−0.013	−0.016	−0.018	−0.019	−0.019	−0.020	−0.020	−0.020	−0.020
0.8	0.000	−0.002	−0.005	−0.007	−0.009	−0.010	−0.013	−0.016	−0.018	−0.019	−0.019	−0.019	−0.019	−0.019	−0.019
1	0.000	−0.002	−0.003	−0.006	−0.007	−0.008	−0.010	−0.013	−0.015	−0.015	−0.016	−0.016	−0.016	−0.016	−0.016
1.2	0.000	−0.001	−0.002	−0.004	−0.005	−0.006	−0.007	−0.009	−0.011	−0.012	−0.012	−0.012	−0.013	−0.013	−0.013
1.4	0.000	−0.001	−0.002	−0.003	−0.003	−0.004	−0.005	−0.007	−0.008	−0.009	−0.009	−0.009	−0.010	−0.010	−0.010
1.5	0.000	−0.001	−0.001	−0.002	−0.003	−0.003	−0.004	−0.006	−0.007	−0.008	−0.008	−0.008	−0.008	−0.009	−0.009
1.6	0.000	−0.001	−0.001	−0.002	−0.002	−0.003	−0.004	−0.005	−0.006	−0.007	−0.007	−0.007	−0.007	−0.008	−0.008
1.8	0.000	−0.000	−0.001	−0.001	−0.002	−0.002	−0.003	−0.004	−0.005	−0.005	−0.005	−0.006	−0.006	−0.006	−0.006
2	0.000	−0.000	−0.001	−0.001	−0.001	−0.002	−0.002	−0.003	−0.003	−0.004	−0.004	−0.004	−0.005	−0.005	−0.005
2.5	0.000	−0.000	−0.000	−0.000	−0.001	−0.001	−0.001	−0.001	−0.002	−0.002	−0.002	−0.002	−0.003	−0.003	−0.003
3	0.000	−0.000	−0.000	−0.000	−0.000	−0.000	−0.001	−0.001	−0.001	−0.001	−0.001	−0.001	−0.002	−0.002	−0.002
4	0.000	−0.000	−0.000	−0.000	−0.000	−0.000	−0.000	−0.000	−0.000	−0.000	−0.001	−0.001	−0.001	−0.001	−0.001
5	0.000	−0.000	−0.000	−0.000	−0.000	−0.000	−0.000	−0.000	−0.000	−0.000	−0.000	−0.000	−0.000	−0.000	−0.000
10	0.000	−0.000	−0.000	−0.000	−0.000	−0.000	−0.000	−0.000	−0.000	−0.000	−0.000	−0.000	−0.000	−0.000	−0.000
15	0.000	−0.000	−0.000	−0.000	−0.000	−0.000	−0.000	−0.000	−0.000	−0.000	−0.000	−0.000	−0.000	−0.000	−0.000
20	0.000	−0.000	−0.000	−0.000	−0.000	−0.000	−0.000	−0.000	−0.000	−0.000	−0.000	−0.000	−0.000	−0.000	−0.000
50	0.000	−0.000	−0.000	−0.000	−0.000	−0.000	−0.000	−0.000	−0.000	−0.000	−0.000	−0.000	−0.000	−0.000	−0.000

(After Giroud)

Table XIV. Coefficient M_3

Depth z/L for the corner, $2z/L$ for the centre	B/L: 0	0.1	0.2	1/3	0.4	0.5	2/3	1	1.5	2	2.5	3	5	10	∞
0	0.000	∞	∞	∞	∞	∞	∞	∞	∞	∞	∞	∞	∞	∞	∞
0.2	0.000	0.013	0.017	0.012	0.008	0.002	−0.006	−0.015	−0.019	−0.020	−0.021	−0.021	−0.021	−0.021	−0.021
0.4	0.000	−0.006	−0.012	−0.020	−0.024	−0.028	−0.035	−0.042	−0.046	−0.047	−0.047	−0.048	−0.048	−0.048	−0.048
0.5	0.000	−0.006	−0.012	−0.019	−0.022	−0.026	−0.031	−0.037	−0.041	−0.042	−0.042	−0.043	−0.043	−0.043	−0.043
0.6	0.000	−0.005	−0.010	−0.015	−0.018	−0.021	−0.026	−0.031	−0.034	−0.035	−0.036	−0.036	−0.036	−0.036	−0.036
0.8	0.000	−0.003	−0.006	−0.009	−0.011	−0.013	−0.016	−0.020	−0.022	−0.023	−0.023	−0.023	−0.024	−0.024	−0.024
1	0.000	−0.002	−0.003	−0.005	−0.006	−0.008	−0.009	−0.012	−0.014	−0.014	−0.015	−0.015	−0.015	−0.015	−0.015
1.2	0.000	−0.001	−0.002	−0.003	−0.004	−0.004	−0.006	−0.007	−0.009	−0.009	−0.009	−0.010	−0.010	−0.001	−0.010
1.4	0.000	−0.001	−0.001	−0.002	−0.002	−0.003	−0.003	−0.004	−0.005	−0.006	−0.006	−0.006	−0.006	−0.006	−0.006
1.5	0.000	−0.001	−0.001	−0.001	−0.002	−0.002	−0.003	−0.004	−0.004	−0.005	−0.005	−0.005	−0.005	−0.005	−0.005
1.6	0.000	−0.000	−0.001	−0.001	−0.001	−0.002	−0.002	−0.003	−0.004	−0.004	−0.004	−0.004	−0.004	−0.004	−0.004
1.8	0.000	−0.000	−0.000	−0.001	−0.001	−0.001	−0.001	−0.002	−0.002	−0.003	−0.003	−0.003	−0.003	−0.003	−0.003
2	0.000	−0.000	−0.000	−0.000	−0.001	−0.001	−0.001	−0.001	−0.002	−0.002	−0.002	−0.002	−0.002	−0.002	−0.002
2.5	0.000	−0.000	−0.000	−0.000	−0.000	−0.000	−0.000	−0.000	−0.001	−0.001	−0.001	−0.001	−0.001	−0.001	−0.001
3	0.000	−0.000	−0.000	−0.000	−0.000	−0.000	−0.000	−0.000	−0.000	−0.000	−0.000	−0.000	−0.000	−0.000	−0.000
4	0.000	−0.000	−0.000	−0.000	−0.000	−0.000	−0.000	−0.000	−0.000	−0.000	−0.000	−0.000	−0.000	−0.000	−0.000
5	0.000	−0.000	−0.000	−0.000	−0.000	−0.000	−0.000	−0.000	−0.000	−0.000	−0.000	−0.000	−0.000	−0.000	−0.000
10	0.000	−0.000	−0.000	−0.000	−0.000	−0.000	−0.000	−0.000	−0.000	−0.000	−0.000	−0.000	−0.000	−0.000	−0.000
15	0.000	−0.000	−0.000	−0.000	−0.000	−0.000	−0.000	−0.000	−0.000	−0.000	−0.000	−0.000	−0.000	−0.000	−0.000
20	0.000	−0.000	−0.000	−0.000	−0.000	−0.000	−0.000	−0.000	−0.000	−0.000	−0.000	−0.000	−0.000	−0.000	−0.000
50	0.000	−0.000	−0.000	−0.000	−0.000	−0.000	−0.000	−0.000	−0.000	−0.000	−0.000	−0.000	−0.000	−0.000	−0.000

(After Giroud)

Table XV. Coefficient M_5

Depth z/L for the corner, $2z/L$ for the centre	B/L: 0	0.1	0.2	1/3	0.4	0.5	2/3	1	1.5	2	2.5	3	5	10	∞
0	0.000	∞	∞	∞	∞	∞	∞	∞	∞	∞	∞	∞	∞	∞	∞
0.2	0.000	0.003	0.015	0.031	0.036	0.042	0.046	0.045	0.040	0.036	0.034	0.032	0.030	0.029	0.028
0.4	0.000	0.000	0.001	0.004	0.006	0.007	0.008	0.006	0.002	−0.002	−0.004	−0.005	−0.008	−0.009	−0.009
0.5	0.000	0.000	0.001	0.002	0.002	0.003	0.003	0.001	−0.004	−0.007	−0.009	−0.010	−0.013	−0.014	−0.014
0.6	0.000	0.000	0.000	0.001	0.001	0.001	0.000	−0.002	−0.006	−0.009	−0.011	−0.012	−0.014	−0.015	−0.016
0.8	0.000	−0.000	−0.000	−0.000	−0.000	−0.000	−0.001	−0.003	−0.006	−0.009	−0.010	−0.012	−0.014	−0.015	−0.015
1	0.000	−0.000	−0.000	−0.000	−0.000	−0.000	−0.001	−0.002	−0.005	−0.007	−0.009	−0.010	−0.012	−0.013	−0.013
1.2	0.000	−0.000	−0.000	−0.000	−0.000	−0.000	−0.001	−0.002	−0.004	−0.005	−0.007	−0.008	−0.010	−0.011	−0.011
1.4	0.000	−0.000	−0.000	−0.000	−0.000	−0.000	−0.000	−0.001	−0.003	−0.004	−0.005	−0.006	−0.008	−0.009	−0.009
1.5	0.000	−0.000	−0.000	−0.000	−0.000	−0.000	−0.000	−0.001	−0.002	−0.003	−0.004	−0.005	−0.007	−0.008	−0.008
1.6	0.000	−0.000	−0.000	−0.000	−0.000	−0.000	−0.000	−0.001	−0.002	−0.003	−0.004	−0.005	−0.006	−0.007	−0.008
1.8	0.000	−0.000	−0.000	−0.000	−0.000	−0.000	−0.000	−0.001	−0.001	−0.002	−0.003	−0.004	−0.005	−0.006	−0.006
2	0.000	−0.000	−0.000	−0.000	−0.000	−0.000	−0.000	−0.000	−0.001	−0.002	−0.002	−0.003	−0.004	−0.005	−0.005
2.5	0.000	−0.000	−0.000	−0.000	−0.000	−0.000	−0.000	−0.000	−0.000	−0.001	−0.001	−0.002	−0.003	−0.003	−0.004
3	0.000	−0.000	−0.000	−0.000	−0.000	−0.000	−0.000	−0.000	−0.000	−0.000	−0.001	−0.001	−0.002	−0.002	−0.003
4	0.000	−0.000	−0.000	−0.000	−0.000	−0.000	−0.000	−0.000	−0.000	−0.000	−0.000	−0.000	−0.001	−0.001	−0.002
5	0.000	−0.000	−0.000	−0.000	−0.000	−0.000	−0.000	−0.000	−0.000	−0.000	−0.000	−0.000	−0.000	−0.001	−0.001
10	0.000	−0.000	−0.000	−0.000	−0.000	−0.000	−0.000	−0.000	−0.000	−0.000	−0.000	−0.000	−0.000	−0.000	−0.000
15	0.000	−0.000	−0.000	−0.000	−0.000	−0.000	−0.000	−0.000	−0.000	−0.000	−0.000	−0.000	−0.000	−0.000	−0.000
20	0.000	−0.000	−0.000	−0.000	−0.000	−0.000	−0.000	−0.000	−0.000	−0.000	−0.000	−0.000	−0.000	−0.000	−0.000
50	0.000	−0.000	−0.000	−0.000	−0.000	−0.000	−0.000	−0.000	−0.000	0.000	−0.000	−0.000	−0.000	−0.000	−0.000

(After Giroud)

Table XVI. Coefficient M'_2

Depth z/L	B/L = 0	0.1	0.2	1/3	0.4	0.5	2/3	1	1.5	2	2.5	3	5	10	∞
0	0.000	0.161	0.115	0.077	0.063	0.048	0.031	0.015	0.006	0.003	0.002	0.001	0.000	0.000	0.000
0.2	0.000	0.026	0.037	0.036	0.033	0.028	0.021	0.011	0.005	0.002	0.001	0.001	0.000	0.000	0.000
0.4	0.000	0.007	0.012	0.015	0.015	0.015	0.012	0.008	0.004	0.002	0.001	0.001	0.000	0.000	0.000
0.5	0.000	0.004	0.008	0.010	0.010	0.010	0.009	0.006	0.003	0.002	0.001	0.001	0.000	0.000	0.000
0.6	0.000	0.003	0.005	0.007	0.007	0.008	0.007	0.005	0.003	0.002	0.001	0.001	0.000	0.000	0.000
0.8	0.000	0.001	0.002	0.003	0.004	0.004	0.004	0.003	0.002	0.001	0.001	0.001	0.000	0.000	0.000
1	0.000	0.001	0.001	0.002	0.002	0.002	0.002	0.002	0.002	0.001	0.001	0.000	0.000	0.000	0.000
1.2	0.000	0.000	0.001	0.001	0.001	0.001	0.002	0.002	0.001	0.001	0.001	0.000	0.000	0.000	0.000
1.4	0.000	0.000	0.000	0.001	0.001	0.001	0.001	0.001	0.001	0.001	0.001	0.000	0.000	0.000	0.000
1.5	0.000	0.000	0.000	0.000	0.001	0.001	0.001	0.001	0.001	0.001	0.000	0.000	0.000	0.000	0.000
1.6	0.000	0.000	0.000	0.000	0.000	0.001	0.001	0.001	0.001	0.001	0.000	0.000	0.000	0.000	0.000
1.8	0.000	0.000	0.000	0.000	0.000	0.000	0.000	0.001	0.001	0.000	0.000	0.000	0.000	0.000	0.000
2	0.000	0.000	0.000	0.000	0.000	0.000	0.000	0.000	0.000	0.000	0.000	0.000	0.000	0.000	0.000
2.5	0.000	0.000	0.000	0.000	0.000	0.000	0.000	0.000	0.000	0.000	0.000	0.000	0.000	0.000	0.000
3	0.000	0.000	0.000	0.000	0.000	0.000	0.000	0.000	0.000	0.000	0.000	0.000	0.000	0.000	0.000
4	0.000	0.000	0.000	0.000	0.000	0.000	0.000	0.000	0.000	0.000	0.000	0.000	0.000	0.000	0.000
5	0.000	0.000	0.000	0.000	0.000	0.000	0.000	0.000	0.000	0.000	0.000	0.000	0.000	0.000	0.000
10	0.000	0.000	0.000	0.000	0.000	0.000	0.000	0.000	0.000	0.000	0.000	0.000	0.000	0.000	0.000
15	0.000	0.000	0.000	0.000	0.000	0.000	0.000	0.000	0.000	0.000	0.000	0.000	0.000	0.000	0.000
20	0.000	0.000	0.000	0.000	0.000	0.000	0.000	0.000	0.000	0.000	0.000	0.000	0.000	0.000	0.000
50	0.000	0.000	0.000	0.000	−0.000	0.000	0.000	0.000	0.000	0.000	0.000	0.000	0.000	0.000	0.000

(After Giroud)

Table XVII. Coefficient M'_3

Depth: z/L for the corner, $2z/L$ for the centre	B/L = 0	0.1	0.2	1/3	0.4	0.5	2/3	1	1.5	2	2.5	3	5	10	∞
0	0.000	0.080	0.043	0.017	0.009	0.001	−0.006	−0.009	−0.007	−0.005	−0.003	−0.001	−0.001	−0.000	0.000
0.2	0.000	0.004	0.005	0.001	−0.001	−0.003	−0.006	−0.007	−0.006	−0.004	−0.003	−0.002	−0.001	−0.000	0.000
0.4	0.000	−0.000	−0.001	−0.002	−0.003	−0.003	−0.004	−0.005	−0.004	−0.003	−0.003	−0.002	−0.001	−0.000	0.000
0.5	0.000	−0.001	−0.001	−0.002	−0.003	−0.003	−0.004	−0.004	−0.004	−0.003	−0.002	−0.002	−0.001	−0.000	0.000
0.6	0.000	−0.001	−0.001	−0.002	−0.002	−0.003	−0.003	−0.004	−0.003	−0.003	−0.002	−0.002	−0.001	−0.000	0.000
0.8	0.000	−0.000	−0.001	−0.001	−0.002	−0.002	−0.002	−0.003	−0.003	−0.002	−0.002	−0.001	−0.001	−0.000	0.000
1	0.000	−0.000	−0.001	−0.001	−0.001	−0.001	−0.002	−0.002	−0.002	−0.002	−0.002	−0.001	−0.001	−0.000	0.000
1.2	0.000	−0.000	−0.000	−0.001	−0.001	−0.001	−0.001	−0.002	−0.002	−0.002	−0.001	−0.001	−0.001	−0.000	0.000
1.4	0.000	−0.000	−0.000	−0.001	−0.001	−0.001	−0.001	−0.001	−0.001	−0.001	−0.001	−0.001	−0.001	−0.000	0.000
1.5	0.000	−0.000	−0.000	−0.000	−0.001	−0.001	−0.001	−0.001	−0.001	−0.001	−0.001	−0.001	−0.001	−0.000	0.000
1.6	0.000	−0.000	−0.000	−0.000	−0.000	−0.001	−0.001	−0.001	−0.001	−0.001	−0.001	−0.001	−0.001	−0.000	0.000
1.8	0.000	−0.000	−0.000	−0.000	−0.000	−0.000	−0.001	−0.001	−0.001	−0.001	−0.001	−0.001	−0.000	−0.000	0.000
2	0.000	−0.000	−0.000	−0.000	−0.000	−0.000	−0.000	−0.001	−0.001	−0.001	−0.001	−0.001	−0.000	−0.000	0.000
2.5	0.000	−0.000	−0.000	−0.000	−0.000	−0.000	−0.000	−0.000	−0.000	−0.000	−0.000	−0.000	−0.000	−0.000	0.000
3	0.000	−0.000	−0.000	−0.000	−0.000	−0.000	−0.000	−0.000	−0.000	−0.000	−0.000	−0.000	−0.000	−0.000	0.000
4	0.000	−0.000	−0.000	−0.000	−0.000	−0.000	−0.000	−0.000	−0.000	−0.000	−0.000	−0.000	−0.000	−0.000	−0.000
5	0.000	−0.000	−0.000	−0.000	−0.000	−0.000	−0.000	−0.000	−0.000	−0.000	−0.000	−0.000	−0.000	−0.000	0.000
10	0.000	−0.000	−0.000	−0.000	−0.000	−0.000	−0.000	−0.000	−0.000	−0.000	−0.000	−0.000	−0.000	−0.000	0.000
15	0.000	−0.000	−0.000	−0.000	−0.000	−0.000	−0.000	−0.000	−0.000	−0.000	−0.000	−0.000	−0.000	−0.000	0.000
20	0.000	−0.000	−0.000	−0.000	−0.000	−0.000	−0.000	−0.000	−0.000	−0.000	−0.000	−0.000	−0.000	−0.000	0.000
50	0.000	−0.000	−0.000	−0.000	−0.000	−0.000	−0.000	−0.000	−0.000	−0.000	−0.000	−0.000	−0.000	−0.000	0.000

(After Giroud)

Table XVIII. Coefficient M'_5

Depth z/L for the corner, $2z/L$ for the centre	B/L = 0	0.1	0.2	1/3	0.4	0.5	2/3	1	1.5	2	2.5	3	5	10	∞
0	0.000	∞	∞	∞	∞	∞	∞	∞	∞	∞	∞	∞	∞	∞	∞
0.2	0.000	0.016	0.028	0.038	0.041	0.044	0.045	0.043	0.038	0.035	0.033	0.032	0.030	0.029	0.028
0.4	0.000	0.002	0.004	0.006	0.006	0.006	0.006	0.003	−0.000	−0.003	−0.005	−0.006	−0.008	−0.009	−0.009
0.5	0.000	0.001	0.001	0.001	0.001	0.001	0.000	−0.002	−0.006	−0.008	−0.010	−0.011	−0.013	−0.014	−0.014
0.6	0.000	−0.000	−0.000	−0.001	−0.001	−0.002	−0.003	−0.005	−0.008	−0.010	−0.012	−0.013	−0.014	−0.015	−0.016
0.8	0.000	−0.001	−0.001	−0.002	−0.002	−0.003	−0.004	−0.006	−0.009	−0.010	−0.012	−0.012	−0.014	−0.015	−0.015
1	0.000	−0.001	−0.001	−0.002	−0.002	−0.003	−0.004	−0.005	−0.007	−0.009	−0.010	−0.011	−0.012	−0.013	−0.013
1.2	0.000	−0.000	−0.001	−0.002	−0.002	−0.002	−0.003	−0.004	−0.006	−0.007	−0.008	−0.009	−0.010	−0.011	−0.011
1.4	0.000	−0.000	−0.001	−0.001	−0.001	−0.002	−0.002	−0.003	−0.005	−0.006	−0.006	−0.007	−0.008	−0.009	−0.009
1.5	0.000	−0.000	−0.001	−0.001	−0.001	−0.002	−0.002	−0.003	−0.004	−0.005	−0.006	−0.006	−0.007	−0.008	−0.008
1.6	0.000	−0.000	−0.001	−0.001	−0.001	−0.001	−0.002	−0.003	−0.004	−0.004	−0.005	−0.006	−0.007	−0.007	−0.008
1.8	0.000	−0.000	−0.000	−0.001	−0.001	−0.001	−0.001	−0.002	−0.003	−0.004	−0.004	−0.005	−0.005	−0.006	−0.006
2	0.000	−0.000	−0.000	−0.001	−0.001	−0.001	−0.001	−0.002	−0.002	−0.003	−0.003	−0.004	−0.004	−0.005	−0.005
2.5	0.000	−0.000	−0.000	−0.000	−0.000	−0.001	−0.001	−0.001	−0.001	−0.002	−0.002	−0.002	−0.003	−0.003	−0.004
3	0.000	−0.000	−0.000	−0.000	−0.000	−0.000	−0.000	−0.001	−0.001	−0.001	−0.001	−0.002	−0.002	−0.002	−0.003
4	0.000	−0.000	−0.000	−0.000	−0.000	−0.000	−0.000	−0.000	−0.000	−0.001	−0.001	−0.001	−0.001	−0.001	−0.002
5	0.000	−0.000	−0.000	−0.000	−0.000	−0.000	−0.000	−0.000	−0.000	−0.000	−0.000	−0.000	−0.001	−0.001	−0.001
10	0.000	−0.000	−0.000	−0.000	−0.000	−0.000	−0.000	−0.000	−0.000	−0.000	−0.000	−0.000	−0.000	−0.000	−0.000
15	0.000	−0.000	−0.000	−0.000	−0.000	−0.000	−0.000	−0.000	−0.000	−0.000	−0.000	−0.000	−0.000	−0.000	−0.000
20	0.000	−0.000	−0.000	−0.000	−0.000	−0.000	−0.000	−0.000	−0.000	−0.000	−0.000	−0.000	−0.000	−0.000	−0.000
50	0.000	−0.000	−0.000	−0.000	−0.000	−0.000	−0.000	−0.000	−0.000	−0.000	−0.000	−0.000	−0.000	−0.000	−0.000

(After Giroud)

Table XIX. Coefficient N_2

Depth z/L	B/L = 0	0.1	0.2	1/3	0.4	0.5	2/3	1	1.5	2	2.5	3	5	10	∞
0	0.000	0.250	0.250	0.250	0.250	0.250	0.250	0.250	0.250	0.250	0.250	0.250	0.250	0.250	0.250
0.2	0.000	0.008	0.034	0.067	0.078	0.090	0.101	0.110	0.114	0.114	0.115	0.115	0.115	0.115	0.115
0.4	0.000	0.001	0.006	0.018	0.024	0.033	0.044	0.056	0.062	0.063	0.063	0.063	0.063	0.063	0.062
0.5	0.000	0.000	0.003	0.010	0.014	0.020	0.029	0.040	0.045	0.047	0.048	0.048	0.048	0.048	0.048
0.6	0.000	0.000	0.002	0.006	0.008	0.013	0.019	0.028	0.034	0.036	0.036	0.037	0.037	0.037	0.037
0.8	0.000	0.000	0.000	0.002	0.003	0.005	0.009	0.015	0.019	0.021	0.022	0.022	0.023	0.023	0.023
1	0.000	0.000	0.000	0.001	0.001	0.002	0.004	0.008	0.011	0.013	0.014	0.014	0.015	0.015	0.015
1.2	0.000	0.000	0.000	0.000	0.001	0.001	0.002	0.004	0.007	0.008	0.009	0.009	0.010	0.010	0.010
1.4	0.000	0.000	0.000	0.000	0.000	0.000	0.001	0.002	0.004	0.005	0.006	0.006	0.007	0.007	0.007
1.5	0.000	0.000	0.000	0.000	0.000	0.000	0.001	0.002	0.003	0.004	0.005	0.005	0.006	0.006	0.006
1.6	0.000	0.000	0.000	0.000	0.000	0.000	0.001	0.001	0.003	0.004	0.004	0.004	0.005	0.005	0.005
1.8	0.000	0.000	0.000	0.000	0.000	0.000	0.000	0.001	0.002	0.002	0.003	0.003	0.004	0.004	0.004
2	0.000	0.000	0.000	0.000	0.000	0.000	0.000	0.001	0.001	0.002	0.002	0.002	0.003	0.003	0.003
2.5	0.000	0.000	0.000	0.000	0.000	0.000	0.000	0.000	0.000	0.001	0.001	0.001	0.001	0.001	0.002
3	0.000	0.000	0.000	0.000	0.000	0.000	0.000	0.000	0.000	0.000	0.000	0.001	0.001	0.001	0.001
4	0.000	0.000	0.000	0.000	0.000	0.000	0.000	0.000	0.000	0.000	0.000	0.000	0.000	0.000	0.000
5	0.000	0.000	0.000	0.000	0.000	0.000	0.000	0.000	0.000	0.000	0.000	0.000	0.000	0.000	0.000
10	0.000	0.000	0.000	0.000	−0.000	0.000	0.000	0.000	0.000	0.000	0.000	0.000	0.000	0.000	0.000
15	0.000	0.000	0.000	0.000	0.000	−0.000	0.000	−0.000	0.000	0.000	0.000	0.000	0.000	0.000	0.000
20	0.000	0.000	0.000	0.000	0.000	0.000	0.000	0.000	0.000	0.000	0.000	0.000	0.000	0.000	0.000
50	0.000	0.000	0.000	0.000	0.000	0.000	−0.000	0.000	0.000	−0.000	0.000	0.000	0.000	0.000	0.000

(After Giroud)

Table XX. Coefficient N_2'

Depth z/L	B/L: 0	0.1	0.2	1/3	0.4	0.5	2/3	1	1.5	2	2.5	3	5	10	∞
0	0.000	0.039	0.135	0.173	0.187	0.202	0.219	0.235	0.244	0.247	0.248	0.249	0.250	0.250	0.250
0.2	0.000	0.022	0.042	0.062	0.070	0.079	0.091	0.103	0.110	0.112	0.113	0.114	0.115	0.115	0.115
0.4	0.000	0.009	0.018	0.028	0.033	0.038	0.045	0.054	0.059	0.061	0.062	0.063	0.063	0.063	0.063
0.5	0.000	0.006	0.013	0.020	0.023	0.027	0.033	0.040	0.044	0.046	0.047	0.047	0.048	0.048	0.048
0.6	0.000	0.005	0.009	0.014	0.017	0.020	0.024	0.029	0.033	0.035	0.036	0.036	0.037	0.037	0.037
0.8	0.000	0.002	0.005	0.008	0.009	0.011	0.013	0.017	0.020	0.021	0.022	0.022	0.023	0.023	0.023
1	0.000	0.001	0.003	0.004	0.005	0.006	0.008	0.010	0.012	0.013	0.014	0.014	0.015	0.015	0.015
1.2	0.000	0.001	0.002	0.003	0.003	0.004	0.005	0.006	0.008	0.009	0.009	0.009	0.010	0.010	0.010
1.4	0.000	0.000	0.001	0.002	0.002	0.002	0.003	0.004	0.005	0.006	0.006	0.006	0.007	0.007	0.007
1.5	0.000	0.000	0.001	0.001	0.002	0.002	0.002	0.003	0.004	0.005	0.005	0.005	0.006	0.006	0.006
1.6	0.000	0.000	0.001	0.001	0.001	0.001	0.002	0.003	0.003	0.004	0.004	0.004	0.005	0.005	0.005
1.8	0.000	0.000	0.000	0.001	0.001	0.001	0.001	0.002	0.002	0.003	0.003	0.003	0.004	0.004	0.004
2	0.000	0.000	0.000	0.000	0.001	0.001	0.001	0.001	0.002	0.002	0.002	0.002	0.003	0.003	0.003
2.5	0.000	0.000	0.000	0.000	0.000	0.000	0.000	0.001	0.001	0.001	0.001	0.001	0.001	0.001	0.002
3	0.000	0.000	0.000	0.000	0.000	0.000	0.000	0.000	0.000	0.001	0.001	0.001	0.001	0.001	0.001
4	0.000	0.000	0.000	0.000	0.000	0.000	0.000	0.000	0.000	0.000	0.000	0.000	0.000	0.000	0.000
5	0.000	0.000	0.000	0.000	0.000	0.000	0.000	0.000	0.000	0.000	0.000	0.000	0.000	0.000	0.000
10	0.000	0.000	0.000	0.000	0.000	0.000	0.000	0.000	0.000	0.000	0.000	0.000	0.000	0.000	0.000
15	0.000	0.000	0.000	0.000	0.000	0.000	0.000	0.000	0.000	0.000	0.000	0.000	0.000	0.000	−0.000
20	0.000	0.000	0.000	0.000	0.000	0.000	0.000	0.000	0.000	0.000	0.000	0.000	0.000	0.000	−0.000
50	0.000	0.000	0.000	0.000	0.000	0.000	−0.000	0.000	0.000	−0.000	0.000	0.000	0.000	0.000	−0.000

(After Giroud)

AUTHOR INDEX

SUBJECT INDEX

Made in United States
Orlando, FL
04 November 2026

90200666R00120